STUDIES ON VOLTAIRE AND THE EIGHTEENTH CENTURY

SVEC
2008:09

Manuscripts should be prepared in accordance with the *SVEC* style sheet, available on request and at the Voltaire Foundation website (www.voltaire.ox.ac.uk). One paper copy should be submitted to the *SVEC* general editor at the Voltaire Foundation, 99 Banbury Road, Oxford OX2 6JX, UK; an electronic version, with a summary of about 750 words, should be sent to jonathan.mallinson@trinity.ox.ac.uk.

Diderot: natural philosopher

KURT BALLSTADT

SVEC
2008:09

VOLTAIRE FOUNDATION
OXFORD
2008

ISBN 978 0 7294 0948 3
ISSN 0435-2866

Voltaire Foundation
99 Banbury Road
Oxford OX2 6JX, UK

A catalogue record for this book
is available from the British Library

The correct reference for this volume is
SVEC 2008:09

This series is available on annual subscription

For further information about *SVEC*
and other Voltaire Foundation publications see
www.voltaire.ox.ac.uk, or email svec@voltaire.ox.ac.uk

This book is printed on acid-free paper

Printed in UK by the Alden Press, Oxfordshire

For Sarah and Alexandra

Contents

List of abbreviations

OC	Unless otherwise stated, all references to Diderot works have been taken from *Œuvres complètes* (édition chronologique), éd. Roger Lewinter (Paris, 1963-1973)
Encyclopédie	*Encyclopédie, ou, Dictionnaire raisonné des sciences, des arts et des métiers* (Paris, Briasson *et al*, 1751-1765)

Introduction

In the *Encyclopédie* article 'Eclectisme' Denis Diderot outlines a philosophy that is marked by both a grandiose vision and a remarkable sense of humility. He paints a vivid and rather passionate picture of an approach to life that seeks nothing less than a transformation of the self; a transformation whereby one becomes a 'disciple of humanity'. This is achieved, at least in part, through a process of continual learning that embraces all manner of disciplines, and that encompasses a tremendously wide variety of perspectives.[1] There is undoubtedly an autobiographical element to this definition, for, perhaps appropriate to one who spent the better part of his career editing and compiling a gargantuan project like the *Encyclopédie*, Diderot himself possessed an unusually broad range of passions, interests and aptitudes. Perhaps even more impressively, he had the perceptiveness and intellectual fibre to offer novel insights and intelligent criticisms in many of the fields that captured his imagination. Of course despite, or perhaps because of this tremendous versatility, certain aspects of Diderot's intellectual career have received more attention than others, and consequently some realms of his interest still remain understudied and underdeveloped. One such area is the *philosophe*'s interest in the natural sciences.

To date the only truly comprehensive attempt to appraise Diderot's natural philosophy has been Jean Mayer's *Diderot, homme de science*, written in 1959.[2] Much has happened since 1959 however, not only within the realm of Diderot studies but, perhaps more significantly, within the realms of the history and philosophy of science. Since the publication of Mayer's work many new vistas have been opened up within these disciplines, new models have appeared for examining the form and content of a given scientific oeuvre, and fresh perspectives on the natural philosophical landscape of the eighteenth century have been developed. The last few decades have seen scholars form an unprecedented appreciation for the complexities of the relationship between core and periphery, between professional and amateur science, and between science and literature. If Diderot's natural philosophical thought is to be taken seriously, then the fruit of these developments in the history and philosophy of science must be considered to impact upon an evaluation of his scientific career.

This is not to say, of course, that there has been no recent expression of interest in Diderot's natural philosophy. Appropriate to the general burgeoning of the history of science that commenced in the early 1960s,

1. Denis Diderot, 'Eclectisme', *Encyclopédie*, vol.5 (1755), p.270.
2. Jean Mayer, *Diderot, homme de science* (Rennes, 1959).

a number of pieces have appeared since then regarding scientific elements of the *philosophe*'s thought. These studies have taken the form of various articles, chapters or books devoted to considering aspects of his canon of natural philosophical work. Most frequently such efforts have tended to focus on Diderot's ideas regarding the sciences of life, a point that is manifested particularly in the amount of attention accorded to the only one of his science-related writings that can truly be called famous, *Le Rêve de D'Alembert.* Although these recent works have had a fairly substantial cumulative impact, they have tended to remain relatively isolated, and none have attempted to emulate the scope of Mayer's overview. Thus, there remains a gap in our elucidation and comprehension of Diderot the scientist – namely that there is no modern survey of his work in this tremendously important realm.

It is precisely this gap that the present book seeks to fill. On the one hand, it strives to draw together the disparate attempts that have appeared over the course of the last forty years to comprehend Diderot's science, and on the other it offers an analysis that is more modern, penetrating and interpretive than Mayer's work. The last of these issues is of special relevance, for while *Diderot, homme de science* is a valuable resource, providing a very useful overview of Diderot's natural philosophical thought, there are elements of its interpretation that must be revised. This is especially the case, as we shall see, with regard to Diderot's mathematics, chemistry and even certain facets of his 'biological' thought. Furthermore Mayer fails to place enough emphasis on the continuity between Diderot's epistemology, his methodology and the actual content of his natural philosophical world-view. As a result, it fails to give Diderot's scientific œuvre the coherence and interconnectedness that it merits.

An example of the perils of neglecting to elucidate fully Diderot's approach to natural philosophy is apparent in works such as Jonathan Israel's *Radical Enlightenment.*[3] The purpose of Israel's book is to argue that the most subversive strains of eighteenth-century thought are almost entirely rooted in seventeenth-century Dutch philosophy, particularly that of Spinoza. As far as Diderot is concerned, Israel's interpretation advances two main points, first, that the *philosophe* is a stereotypical 'radical' and second that his work might be viewed as little more than the culmination of this century-long tradition; that Diderot is merely sitting on the shoulders, as it were, of seventeenth-century Dutch giants. Tellingly, Israel almost entirely neglects to address Diderot as a scientific thinker (although he does pay the seemingly obligatory homage to *Le Rêve de D'Alembert*) and this omission has tremendous repercussions. Granted, the analysis of *Radical Enlightenment* nominally ends in the year 1750 and

3. Jonathan Israel, *Radical Enlightenment: philosophy and the making of modernity 1650-1750* (Oxford, 2002).

by this date, the majority of Diderot's natural philosophical work was yet to be produced. Nevertheless it is maintained here that a thorough examination of Diderot's natural philosophy has the potential to upset Israel's appraisal of his place in the Enlightenment. We shall see in due course that not only is much of Diderot's philosophy predicated upon giving paramount consideration to the study of nature, and that as such he was influenced deeply by a great variety of seventeenth- and eighteenth-century scientists (not just Spinoza and a few Dutch 'radicals'), but that in many respects Diderot may not be such a radical after all. This is not to suggest that the accepted view of Diderot as being unorthodox and in some ways revolutionary needs to be overturned, but simply that this view needs to be nuanced. Seeing him as a brilliant but rather marginalised thinker swimming against the tide, offering up novel and often radical notions, does not really do his natural philosophy justice. He interacted a great deal with 'mainstream' scientists, assimilated many of their ideas and posited his own notions within the context of larger eighteenth-century debates. Diderot was serious about his science, made valid and earnest attempts to grasp and refine (rather than reject) most of the orthodox natural philosophy of his day, and he almost certainly would have objected to being classified as a 'radical'. All of this becomes apparent upon an examination of his scientific oeuvre that stresses its wholeness, unity of purpose and desire for consistency and coherence.

Of course, many Diderot scholars will note here that any attempt to impress a sense of unity and interconnectedness upon his thought is potentially precarious and perhaps to be avoided entirely. It has become very much the trend of the last two decades to view Diderot as something of a prototypical postmodernist and to deliberately avoid regarding his work with an eye to any underlying unity. Peter France, for instance, has argued that Diderot is best appreciated when given total autonomy, when we simply take him as he comes and refrain from imposing any sort of overarching interpretation.[4] This general presupposition that the correct way to tackle (and, it should be said, to enjoy) Diderot is to allow him his freedom and try to do no more than examine his work in bits and pieces seems to be reinforced by many commentators. Emita Hill has noted the degree to which almost all of Diderot's works are open to charges of inconsistency and easily subject to manifold interpretations.[5] The essence of this general climate of opinion regarding the *philosophe* and his writings is nicely captured by Jack Undank, who remarks that:

> our fascination with Diderot [...] has less to do with history and philology than with a literary manner that strikes us as remarkably pertinent even in a post-revolutionary, post-idealist, post-modern era, when dispersion, decenterings and

4. Peter France, *Diderot* (Oxford, 1983), p.2-3.

5. Emita Hill, 'Materialism and monsters in *Le Rêve de D'Alembert*', in *Diderot studies* 10 (1968), p.67-68.

suspensive, but also assertive ironics gaze somewhat impatiently into the past for their own reflections.[6]

These studies, allied to a variety of others, have thus promulgated the image – one might even say a highly cherished image – of Diderot as a champion of disorder, a wilful defiler of neatness, one who delights in paradox and is not greatly disturbed by inconsistency or uncertainty.[7] As shall become clear in the pages that follow, these things are not entirely untrue; Diderot did often indulge in paradox and he did posit a view of reality in which absolute certainty had to be regarded as the privilege of a God that did not even exist. Did this mean, however, that he relished disorder? Did Diderot's world-view essentially mock any hopes of something that today would be called science: an ordered, relatively coherent attempt on the part of humankind to grasp the nature of the universe and to regard it from a perspective that incorporates some kind of structure? Moreover is it impossible to perceive in the work and thought of Diderot himself some manner of structure, consistency and wholeness? The answer to all of these questions must be an unequivocal no. While the concerns expressed by France, Hill, Undank and others should not be entirely ignored, there does exist the possibility, indeed the necessity, to step back and take a broader look at Diderot's natural philosophical thought and to try to interpret it as a grand framework.

The underlying assumption of this book is that there is in fact some overarching, integrated effort on Diderot's part to give a coherent and comprehensive account of the world in which humanity finds itself. We witness in his various scientific works certain common threads, many recurring themes and a perpetual drive to make sense of the universe. Unquestionably Diderot had faith in science, both in terms of its practical applications and as the simple act of interpreting nature. The question here must be to address the issues of exactly how this faith was constituted and how it was grounded. Moreover, Diderot regarded reality as a unified whole, and stressed on several occasions that human knowledge of this reality ultimately had to be correspondingly united. This is not to assert that he viewed natural philosophy as a monolithic enterprise, but simply to say that he saw its diversity as a matter of multiple facets rather than as a series of dissociated and competing fields. In other words, Diderot perceived natural philosophy as no more and no less than humankind's attempt to make sense of the whole that is known as the universe – each relevant scientific discipline being but an element of this grand struggle.

6. Jack Undank and Herbert Josephs (ed.), *Diderot: digression and dispersion: a bicentennial tribute* (Lexington, KY, 1984), p.13.

7. Additional examples of what may be called this 'postmodern' take on Diderot may be found in Carol Blum, *Diderot: the virtue of a philosopher* (New York, 1974), James Creech, *Diderot: thresholds of representation* (Columbus, OH, 1986) and Daniel Brewer, *The Discourse of Enlightenment in eighteenth-century France: Diderot and the art of philosophizing* (Cambridge, 1993).

The structure of the work at hand is therefore broken down according to the format of the scientific disciplines in which Diderot was most interested: mathematics, experimental physics, chemistry, natural history and medicine. This multidisciplinary structure has been adopted partly due to its convenience, but for several other reasons as well. First, it offers the opportunity to devote unprecedented attention to certain aspects of Diderot's natural philosophical thought that have tended to be either misconstrued or underestimated, or both. Most notable here, as has already been suggested with regard to Mayer's book, are mathematics and chemistry. The second reason for proceeding along such multidisciplinary lines is that this probably provides the best means to illustrate as explicitly as possible the fact that while each facet or field might bear semi-autonomous treatment, there are in fact areas of profound overlapping. Ultimately none of the disciplines can be extricated from the complex fabric that is comprised by their interdependence and continually evolving interplay. It is hoped that by treating each facet on its own, while simultaneously referring to its linkage with all of the other disciplines, the true nature of their interpenetration and connection will be properly made manifest.

It should also be stressed that in adopting an appropriately fuzzied but nonetheless multidisciplinary approach, we will in a sense be following in the footsteps of Diderot himself, for in *Plan d'une université* (1775), a pedagogical scheme conceived for the benefit of Catherine II, the *philosophe* laid out his conception of the various aspects of natural philosophy and the manner in which they should be considered. This work neatly highlights the fact that for Diderot there were necessary connections between the sciences, and an equally necessary order in which they should be regarded and taught.

Of course, *Plan d'une université* has been relatively neglected by Diderot scholars; most attempts at understanding his conception of the sciences have traditionally been directed towards the article 'Encyclopédie'. The pages that follow will discuss both of these works, as each is seminal to Diderot's natural philosophy, but it is the former that really provides the clearest means by which to infiltrate and unpack Diderot's ideas as to how the sciences fit together to form a loose but nonetheless real system of human knowledge.

In a sense then, *Plan d'une université* provides the inspiration for the organisational principle of this study and there are a number of compelling reasons for regarding it in such a fashion. First, it was composed relatively late in Diderot's career and therefore, unlike the article 'Encyclopédie', represents a more mature outlook, the vision at which Diderot had arrived after most of his natural philosophical meditation had been completed. Diderot's ideas, like those of anyone else, evolved significantly over time, and *Plan d'une université* has the air not only of a pedagogical plan but a retrospective of Diderot's own career. It might therefore be seen as an attempt to impose some principle of intelligibility upon the gamut of human knowledge, but with the benefit of hindsight and the weight of experience that Diderot had acquired by the later stages of a career as a natural philosophical thinker. Significantly there is also a sense in which the pedagogical plan sketched out in *Plan d'une université* reflects the process by which Diderot himself became acquainted with the sciences. Apart from symbolic value then, in taking the order presented in *Plan d'une université* as a kind of template for accessing Diderot's conception of the sciences, we are in fact roughly following the chronology of his own natural philosophical development. This point should not be underestimated, for it may well be that Diderot's own experience, his own journey as it were, had an indelible impact upon the way that he thought the sciences should be learned and, crucially, the way that he perceived them to function together. It is therefore no coincidence that here we shall follow the same pattern that Diderot himself delineated in *Plan d'une université*. By this means, it is hoped that a useful frame of reference will be provided with which to contemplate Diderot the natural philosophical thinker.

Thus it is that the first chapter of this book will deal with mathematics, perhaps the most historiographically contentious and problematic of Diderot's passions. Here will be offered an interpretation of the meaning and place of mathematics vis à vis Diderot's natural philosophy and an attendant elucidation of his thoughts regarding his attitude towards the intellectual processes of abstraction and generalisation. In so doing, this chapter will sketch out the absolutely seminal epistemology that underwrites Diderot's entire approach to understanding and ordering nature. The second chapter, experimental physics, will take up where the first leaves off, and devote itself in large part to evaluating Diderot's natural philosophical methodology, with primary reference to *Pensées sur l'interprétation de la nature*. The second part of this chapter will discuss, with reference to the methodology that will have been sketched out, the rough outlines of Diderot's physics, that is, his views regarding the basic nature of matter, force and their interaction. Chapter 3 deals with the impact of chemistry on Diderot's thought, and as such constitutes an analysis of how the discipline lent sophistication, detail and depth to the content of his world-view presented in the previous two chapters.

If these first three chapters present something of a ladder, proceeding from the highest level of abstraction possible (mathematics) down through the more concrete discipline of physics and then onto the level of chemistry, which treats matter and its observed behaviour in an even more intimate fashion, then the fourth chapter must be regarded as taking a slightly lateral step. Natural history is the subject of chapter 4, and with it we will move at once deeper into the heart of reality and yet away from it, back to higher, more removed, general views. Although Diderot was not captivated by natural history per se (that is, he was relatively uninterested in the discipline as it was practised by most of his contemporaries), he was fascinated by considerations regarding the nature and origins of life. Thus, this chapter might equally have been framed as dealing with Diderot's cosmology, for it tackles ideas that are predicated upon his physics and chemistry, but ultimately revolves around his attempts to come to terms with reality on a much grander, more sweeping scale.

The final chapter, which concerns medicine, will attempt to place in perspective the discipline that in many respects represents the pinnacle of Diderot's natural philosophical thought. It shall also be demonstrated however that this branch of natural philosophy, in some ways the culmination of all the rest for Diderot, is perhaps the most problematic and complex of all. Not the least of the reasons for this is that the manner in which Diderot conceived of the 'knowing subject' (itself the object of medical investigation) had tremendous implications for his epistemology. In other words, his epistemology and his physiology were mutually dependent. Therefore the interpretation of Diderot's natural philosophy offered in the pages that follow will by the end have come full circle. Although the chapters on mathematics and medicine each treat epistemological considerations in a slightly different context, by starting with and returning to the issue of how Man knows anything about the world, we will have drawn together two (previously disparate) ends of Diderot's grand natural philosophical vision.

Finally, we might address here a crucial and in many ways obvious fact that will not have escaped the attention of any historian familiar either with Diderot in particular or with the history of science in the eighteenth century – namely, that Diderot himself was not a scientist per se; he did not actually conduct any experiments and, many would argue, he made no substantial theoretical breakthroughs that contributed directly to the course of modern science. Why then should we try to consider him as a natural philosopher at all? One of the keys to answering this question lies in the very nature of the terminology used here, in particular the words 'natural philosopher'. The term 'science' of course did not come into its current state of use (that is, it did not refer to what we today would understand by the term 'science') until the nineteenth century. Thus, in the interests of avoiding anachronism, all of Diderot's contemporaries might be considered 'natural philosophers'. In other words, they should be seen as engaged in an activity that involved coming to grips with the nature of the world, attempting to uncover and interpret its modes of operation, understand (if possible) its underlying causes and, in general, develop a philosophy as to what nature is and how it works. This may not seem very different from the modern usage of the term 'science' but the distinction could be crucial. Was it absolutely seminal for an eighteenth-century thinker to be engaged in active research to be considered a philosopher of nature? Jean Mayer has suggested that it was not. He pointed out, as many have done, that the distinction between amateur and professional scientist in the early modern period can be seen as very ambiguous. Indeed Mayer has even argued that Diderot's status as an 'amateur' placed him at an advantage relative to many of the 'professionals' because he did not face the same pressures and was not constrained by the same weight of institutional, dogmatical, theoretical or systematic considerations that afflicted them.[8] While not necessarily assenting to this reading of the situation, the primary point is that it certainly seems as though there was tremendous scope in the middle of the eighteenth century for an amateur to contemplate publicly available scientific evidence, read and understand the seminal works of the day, meditate deeply, hypothesise, theorise and even develop their own complex and legitimate ideas as to how the world is constituted and how it works. This is exactly what Diderot did. Although he may not have conducted any experiments himself, he kept abreast of the latest developments via the memoirs of the Académie des sciences, he visited a number of scientists in their laboratories and had the opportunity to see them at work. Significantly he was able to comprehend these experiments, the process of drawing conclusions from them and indeed the process of using these conclusions to formulate theories about nature. Moreover Diderot did not simply accept what the professional scientists did, but often drew his own conclusions regarding the evidence that was furnished by their labours, and frequently put this evidence together in accordance with his own philosophical demands, preconceptions and criteria. As we shall see in the following pages, he performed these functions assiduously, thoroughly, with the intention of engaging the best scientific minds of the day, and perhaps most importantly, he performed these functions well.

8. Mayer, *Diderot, homme de science*, p.8.

1. Mathematics

IF Diderot is to be truly appreciated as a natural philosopher, if his work within and enthusiasm for this realm of human endeavour is to undergo a systematic appraisal, then that appraisal must begin with an examination of his interest in mathematics. Such a pursuit may seem a far less glamorous, perhaps even less rewarding pursuit than investigating most other areas of his thought (for example his comparatively famous ideas about the nature and development of life) but it is by no means sterile. Mathematics, and Diderot's ideas about the operation and value of the discipline, was absolutely seminal to his vision of the world.

There are many reasons for turning attention to this realm before any of the other branches of Diderot's thought. Not least of these is that mathematics was one of his first, and greatest, intellectual loves. Before the neo-Baconian empiricism, before the work on medicine and physiology, before the powerful and provocative flights of philosophy that are represented by works such as *Le Rêve de D'Alembert*, there was for Diderot a fascination with geometry, algebra, calculus and probabilities.

It is also worth noting that mathematics was the only branch of eighteenth-century science in which Diderot may be said to have been an actual practitioner. This is undoubtedly due in large part to the circumstances and pressures of his life. Though he offered many insights as to the operation of the world on the physical and chemical levels, for instance, he never had the opportunity to perform experiments himself. Part of the reason for this may simply be lack of will, but it must also be kept in mind that the costs of equipment and material for such undertakings were formidable, and for the majority of his life Diderot was not a rich man. He lacked both the financial ability and the space to pursue properly the experimental sciences of which he became so enamoured. There was also, of course, the added burden of almost twenty-five years spent editing the *Encyclopédie*. It is more than conceivable that the consuming nature of this task precluded him, to a high degree, from partaking in popular amateur activities such as botany. Whatever the reasons, the fact remains that mathematics was a discipline that required very little in the way of apparatus other than one's own mind, and Diderot exercised his accordingly.

The most important reason for taking stock of Diderot's mathematics before any other branch of his scientific thought, however, is that his ideas regarding the nature of the discipline form an epistemological keystone of the complex framework that is his natural philosophy. Mathematics provides the original basis for discussing the interplay in Diderot's thought between empiricism and rationalism. Discussion of mathematics encapsulates discussion of his ideas about abstraction, the construction of knowledge, logic and objectivity. The manner in which Diderot regarded the discipline has tremendous implications for the manner in which he perceived the operation of the universe and the best means of rendering it comprehensible.

Despite the significance of Diderot's interest in mathematics, it is an area of study that has received relatively little attention. It is also an aspect of his philosophical career that, when investigated, has been prone to controversy and confusion. A prime indication of the traditional obscurity of his mathematical work is the fact that Diderot was long regarded as a mathematical philistine. Even André Billy, an early biographer, unquestioningly accepted an anecdote relating to Diderot's stay at the court of Catherine the Great in which the *philosophe* is portrayed as completely ignorant of the discipline. The story, related in Augustus de Morgan's *Budget of Paradoxes*, purports that:

> Diderot paid a visit to the Russian court at the invitation of the Empress. He conversed very freely, and gave the younger members of the Court circle a good deal of lively atheism. The Empress was much amused, but some of her counsellors suggested that it might be desirable to check these expositions of doctrine. The Empress did not like to put a direct muzzle on her guest's tongue, so the following plot was contrived. Diderot was informed that a learned mathematician was in possession of an algebraical demonstration of the existence of God, and would give it him before all the Court, if he desired to hear it. Diderot gladly consented: though the name of the mathematician is not given, it was Euler. He advanced toward Diderot, and said gravely, and in a tone of perfect conviction: 'Monsieur $(a+b^n)/n = x$, donc Dieu existe; Répondez!'
>
> Diderot, to whom algebra was Hebrew, was embarrassed and disconcerted; while peals of laughter rose on all sides. He asked permission to return to France at once, which was granted.[1]

It was nearly a century since the publication of this anecdote before any significant light was shed on Diderot's mathematical exploits. In an article of 1941, Lester Krakeur and Raymond Kruger performed an excellent overview of the *philosophe*'s lifetime of interest in the field. It is an article that provides many useful insights and observations regarding the quality and nature of Diderot's mathematical work. It also praises him for his knowledge of contemporary mathematical issues and methods, and for his novelty with regard to applied mathematics.[2] This general appraisal was echoed by J. L. Coolidge in a short chapter of his work on the mathematics of great amateurs, in which it is noted that even though Diderot 'was not a mathematician of the first rank', he did possess a keen interest in the discipline and performed some 'really stimulating math-

1. Augustus de Morgan, *A Budget of paradoxes* (London, 1872), p.251.

2. Lester Krakeur and Raymond Kruger, 'The Mathematical writings of Diderot', in *Isis*, vol.33, part 2, no.88 (June 1941), p.219-32.

ematical work'.[3] More recently, Robert Morin has provided a glimpse of how mathematics operated for Diderot as an aspect of the imagination.[4] Arguing that Diderot perceived not just mathematics but all of the sciences as an invention of the human mind, Morin's work has gone even further towards redeeming Diderot's interest in the discipline. What is still lacking, however, is an appreciation of how mathematics fits into Diderot's natural philosophy as a whole, of how it underwrites much of what he was to do later in his career, and of the impact this had on his interpretation of nature.

In attempting such a task, we must begin with the manner in which Diderot acquired his mathematical knowledge, for this most certainly had a bearing on the course of his mathematical career. Unfortunately little is known about Diderot's early years, and although scholars have been attempting for some time to assemble a relatively coherent picture of his early intellectual life, it is still a realm very much open to conjecture. Nevertheless it is perhaps more useful to speculate about the possibilities regarding Diderot's initial development than to ignore the issue completely.

Diderot's predilection for the rule of number seems to have begun very early in life. As an older man reflecting back upon his younger days, Diderot proclaimed: 'une des premières choses que mes parents m'apprirent dans l'enfance, c'est l'arithmétique'.[5] The very fact of this remembrance may indicate something of the impression that was made. In any case, his knowledge of, and presumably his fondness for mathematics, was soon to be furthered by his formal education. Although there is relatively little information about his time at school, Diderot is known to have attended the Jesuit college in his hometown of Langres from the age of ten (1723). By his own account a happy and eager student, he excelled in most areas of what Arthur Wilson has described as a non-secular humanistic education, otherwise known as the *ratio studiorum*.[6] Significantly this would have included the basics of arithmetic, and perhaps some fundamental geometry.[7]

From Langres Diderot moved to Paris, where he attended the Jansenist Collège d'Harcourt.[8] He graduated from d'Harcourt with an MA in

3. Julian Lowell Coolidge. *The Mathematics of great amateurs* (Oxford, 1990), p.178-85.

4. See Robert Morin, 'Diderot et l'imagination', in *Annales littéraires de l'Université de Besançon* (Paris, 1987), p.85-101.

5. Herbert Dieckmann, *Inventaire du Fonds Vandeul et Inédits de Diderot* (Geneva, 1951), p.204.

6. Arthur Wilson, *Diderot* (New York, 1972), p.15. For more on the nature of such an education, see *The Jesuits, Ratio Studiorum: plan raisonné et institution des etudes dans la Compagnie de Jesus*, ed. A. Demoustier et D. Julia; translated by L. Albrieux et D. Pralon-Julia (Paris, 1997). See also Roger Chartier *et al.*, *L'Education en France du XVIe au XVIIIe siècle* (Paris, 1976).

7. Lester Crocker, *Diderot: the embattled philosopher* (New York, 1954) p.16.

8. The issue of where Diderot attended college in Paris was long the subject of debate. For many years, scholars cited evidence which seemed to attest to the fact that he actually

philosophy in 1732 and proceeded to study theology, although he appears to have abandoned the course sometime shortly before its completion.[9] In terms of mathematical training, it is unlikely that Diderot would have picked up much here; the Collège d'Harcourt was not renowned for such studies. The possibility remains, however, that Diderot may have spent some time at the Jesuit college of Louis-le-Grand, where he may indeed have gleaned some knowledge of the discipline.[10]

Upon the completion of his formal education, Diderot embarked upon what were to be his 'bohemian years'. Quitting a position as a law clerk and consequently finding himself cut off from the monetary support of his disappointed family, Diderot struggled financially and physically for much of the decade. These years are certain to have been crucial to his intellectual formation. It was primarily out of a desire to pursue intellectual matters that Diderot refused to take up the life of a churchman, lawyer or other 'respectable' profession. Notably, during these years of hand-to-mouth existence, much of the little money that Diderot did make came from work as a tutor of mathematics. This does not, in itself, necessarily mean very much. His charges were children, and the level of expertise required for their tutelage was undoubtedly quite low. Indeed it is possible that Diderot could have been learning along with his pupils, perhaps staying only one step ahead of them at any given time.

That Diderot did possess a relatively extensive knowledge of mathematics at this time is, however, fairly likely. It has been supposed that either during his years at university or shortly thereafter, he was acquainted with a number of prominent Parisian mathematicians. One of these was Dominique Rivard, to whom Diderot refers in both *Réfutation d'Helvétius* and *Plan d'une université*.[11] The other is André Pierre Prémontval, a mathematician who gave private lectures during these

---

attended the Jesuit college of Louis-le-Grand. The matter seems to have been put to rest by Blake Hanna, who argues that he did indeed go to the College d'Harcourt (see Blake Hanna, 'Diderot théologien', in *Revue d'histoire littéraire de France*, vol.78 (1978), p.19-35). The possibility still exists, however, that Diderot had some connection to both schools. His daughter, Mme Vandeul, insisted that Diderot was a school chum of the future Cardinal Bernis, who is most likely to have attended Louis-le-Grand (see Wilson, *Diderot*, p.25-26).

9. Hanna, 'Diderot théologien', p.19-35.

10. Wilson, *Diderot*, p.26.

11. Dominique Rivard (1679-1778) was a professor of philosophy at the Collège de Beauvais from 1735-1770. An early proponent of employing textbooks , he published a very successful manual entitled *Abrégé des élémens de mathématiques*. The popularity of this text may have been due to the fact that its content was rather rudimentary, and Diderot's reference to Rivard in *Plan d'une université* seems to reflect such a judgement, acknowledging that Rivard's text is suitable as an aid to teaching arithmetic and algebra in schools, but that other authors (notably Clairaut and Huygens) are better for elucidating geometry and probabilities. On Rivard see Laurence Brockliss, *French higher education in the seventeenth and eighteenth centuries: a cultural history* (Oxford, 1987), p.190-91 and 382-83. See also Diderot, *Plan d'une université*, *OC*, vol.11, p.775.

decades. Diderot's *Mémoires sur différents sujets de mathématiques* is dedicated to Madame Prémontval, who was herself not unfamiliar with the field of mathematics.[12]

Even more telling is the fact that Diderot was a tutor of mathematics not because he loved children, but because he loved mathematics. By his own account, it was the primary focus of his early intellectual life. In a letter written while in prison at Vincennes, he says that he came to Paris at the age of eighteen and proceeded to indulge in the study of mathematics and *belles lettres*. That the former became for him the subject of tremendous passion is revealed in *Réfutation d'Helvétius*, where Diderot, referring to himself, remarked 'On lui mit entre les mains des cahiers d'arithmétique, d'algèbre et de géométrie qu'il dévora. Entraîné par la suite à des études plus agréables, il se plut à la lecture d'Homère, de Virgile, du Tasse et de Milton, mais revenant toujours aux mathématiques, comme un époux infidèle, las de sa maîtresse, revient de temps en temps à sa femme.'[13]

While it is a difficult (if not ultimately impossible) task to determine exactly how and what Diderot learned of mathematics during the 1720s and 1730s, it is relatively easy to assess the level of competence he had acquired by the early 1740s, for it is at this time that he began to compose and to publish pieces of work involving mathematics. It makes sense to briefly survey this lifetime of mathematical endeavour, for a number of reasons. First, in doing so we will obtain a better sense of Diderot's skill and perhaps therefore go even further towards redeeming his reputation as a mathematician. More importantly though, taking stock of Diderot's canon of mathematical work will allow us to better understand the techniques, problems and issues that interested him most. This is significant, for certain of the issues discussed in Diderot's mathematical works impinge upon the branches of natural philosophy that will be addressed elsewhere in this book, while others introduce us (as they may have introduced Diderot) to some of the methodological and epistemological considerations that would underwrite his world-view.

Diderot's first involvement with a published work came in 1741 when he contributed to a treatise on gnomonics (the science of sundials). The primary author of the work, entitled *Nouveaux Traités de trigonométrie rectiligne et sphérique ... avec un traité de gnomonique*, was Antoine Deparcieux (1703-1768).[14] Although Diderot's contribution is not credited in the published edition, he is said to have worked on the mathematical side of

12. *OC*, vol.2, p.55-56 and p.56, n.1.

13. *Réfutation d'Helvétius*, *OC*, vol.11, p.593.

14. Deparcieux worked not only on celestial mechanics and gnomonics, but on the practical applications of probability theory as well. Several years after the piece on gnomonsis appeared he published the *Essai sur les probabilités de le durée de la vie humaine* (Paris, Frères Guerin, 1746). Significantly, this was a branch of mathematics in which Diderot was to become extremely interested.

the treatise, preparing the general formulae and mathematical tables.[15] This would have required a significant knowledge of trigonometry, geometry, astronomy and algebra, suggesting that even at this early stage of his career, Diderot was a reasonably competent mathematician.

Of greater interest however, is Diderot's 1748 *Mémoires sur différents sujets de mathématiques*. Published by Pissot and Durand, the *Mémoires sur mathématiques* originally appeared in several instalments. This was appropriate to the structure of the work, for it is really a series of five distinct papers on unrelated mathematical subjects. The first and longest of these papers, described by Coolidge as the most important, deals with the science of sound and questions whether preference for certain sounds is arbitrary, or if it is founded in nature.[16] Following Pythagoras, Diderot asserts that perception is the key to musical pleasure and, moreover, that mathematics can and must be applied to musical theory.[17] This involves mathematically describing the vibration of cords and translating the frequency of these vibrations into musical terms. In doing so, Diderot makes reference to the work of a number of mathematicians, including Halley, Flamsteed, Mersenne, Gassendi and Newton. He also makes extensive use of both geometry and logarithms.

The second of the *Mémoires sur mathématiques* is, as was observed by Jean Mayer, the best example of pure mathematics in any of Diderot's work.[18] Fundamentally, it deals with the difference between mechanical and geometric curves. It is a notable item for several reasons. First, it demonstrates Diderot's familiarity and ability to deal with involutes (which have to do with the equivalence of areas and the drawing of transcendental as opposed to algebraic curves).[19] Second, towards the end of the *mémoire* he sketches out a design for a novel type of compass. This was only the first of several occasions on which Diderot would turn his mathematical knowledge towards the development or improvement of mechanical instruments. Perhaps the most remarkable element of the second *mémoire*, however, is that it also represents Diderot's first attempt to come to terms with the ancient problem of squaring the circle, an aspect of his mathematical career to which we shall return.

The third and shortest *mémoire* is essentially a thought experiment. It questions how the tension of a cord fixed at one end and weighted on the

15. Wilson, *Diderot*, p.28-29.

16. Coolidge, *Mathematics of great amateurs*, p.178.

17. *Mémoires sur différents sujets de mathématiques*, *OC*, vol.2.

18. Mayer, *Diderot, homme de science*, p.76.

19. Krakeur and Kruger, 'Mathematical writings', p.222. A metaphor that is often used to define involutes is to say that they are curves traced by the end of a string as it is unwrapped from another curve. Coolidge notes that in employing involutes, Diderot was signalling his approval for Descartes's thesis that geometry should embrace more than that which can be accomplished simply with a ruler and compass. See Coolidge, *Mathematics of great amateurs*, p.181.

other (by an equal and opposite force) can behave in the same manner as a hypothetical cord weighted at both ends. Originally this problem had been proposed by the Italian mathematician and physiologist Giovanni Borelli (1608-1679). As Diderot tells his readers, it was a riddle that had long embarrassed *mécaniciens*. To a certain extent, it embarrassed Diderot as well. Though demonstrating again a more than adequate grasp of the geometry involved (which, in this particular *mémoire*, is admittedly not much), Diderot does not arrive at a definitive solution to the problem. He does propose some ideas however, and urges that the truth of these principles be tested by recourse to experiment.[20]

In the fourth of the *mémoires*, Diderot lays out a plan for a new type of organ that can play any piece of music, yet which is operable by novices in the field of composition. Fundamentally, it is an exercise in mechanics. Although Diderot's description of the object is not as technical as one might wish it to be, this *mémoire* does anticipate the intense fascination that Diderot was to develop regarding the practical (though perhaps, in this case, not that practical!) applications of mathematics. It also reprises the theme of the first *mémoire*, dealing again with mathematics and music, and discussing the mathematically definable 'rapports' that exist between musical notes.

It is the fifth and final *mémoire*, however, that is in many ways the most intriguing of them all, for Diderot uses it as a platform to engage with none other than that paragon of early modern mathematics, Isaac Newton. At issue is calculating the resistance of the air to the motion of a pendulum. Where Newton had used proportionality to arrive at his conclusions, Diderot employed calculus to demonstrate that Newton's results were incorrect. In his attempt to refute this part of Newton's work, Diderot adopts a timid, almost apologetic tone. Nevertheless he was quite correct in his critique of the great mathematician and did succeed in demonstrating that Newton was wrong. It has since been revealed that Diderot's own calculations were incorrect as well, even if slightly more subtle than those of Newton.[21]

This *mémoire* is of interest to the Diderot scholar for at least two reasons. First, it demonstrates conclusively that the *philosophe* was able to use

20. A very Newtonian view to be sure! The other fairly remarkable thing about this *mémoire* is that Diderot directs the reader's attention towards a pair of articles in the upcoming *Encyclopédie*, or, as he refers to it here, the *Dictionnaire universel des sciences et des arts*. 'Voyez,' says Diderot, 'lorsque cet ouvrage paraîtra, les articles CORDE, ou TENSION.' The first of these articles was written, when the time came, by D'Alembert. The second, however, provides a nice demonstration of the heady optimism – perhaps over-optimism – that characterised the early idea of the encyclopaedic project; it never actually appeared.

21. Mayer, *Diderot, homme de science*, p.78. Coolidge maintained that both Newton and Diderot were wide of the mark with respect to this problem on account of the fact that the problem cannot be treated in as summary a fashion as either of them were trying to do. See Coolidge, *Mathematics of great amateurs*, p.184.

calculus. This is significant, for at the time it was still a relatively new branch of mathematics practised only by the very few.[22] Second, the fifth *mémoire* indicates that Diderot was not only familiar with the work of Newton, but that he had the capacity and the mathematical understanding to deal with it in an intelligent and critical manner. This applies not only to Newton's *Opticks*, but to the much more inaccessible *Principia mathematica* as well, to which Diderot makes a number of explicit references in this memoir. In being one of the few that had read and could make sense of Newton's more difficult work, Diderot was certainly distinguished from many of his comrades in the 'Republic of Letters'. Perhaps most notable in this respect is Voltaire, who, although he published a vulgarisation of Newton's work (*Éléments de la philosophie de Newton*, 1739), appears not to have had the same level of mathematical skill as Diderot in terms of appreciating or dealing with the Englishman's masterpiece.[23]

It appears then, that regardless of how he actually acquired his mathematical knowledge, by the late 1740s Diderot had become a very competent mathematician. His love of the discipline, which seems to have been cultivated sometime during the 1730s, was matched by a knowledge of geometry, algebra and calculus. This ability did not go unnoticed by Diderot's contemporaries. *Mémoires sur mathématiques* was acclaimed by the Académie des sciences. Diderot's work was also publicly announced by the *Nouvelles littéraires* of Raynal, and received positive reviews from *Cinq Années litteraires*, the *Journal de Trévoux* and the *Mercure de France*.[24] As the researches of Diana Guiragossian Carr reveal, *Mémoires sur différents sujets de mathématiques* even enjoyed a warm reception in England, where Diderot was praised for being 'very well versed in physics and geometry'.[25]

22. Although it might be said that it was the French who, over the course of the eighteenth century, perfected calculus in a more interesting way than the English, it was still far from being an accessible discipline. It was not until the last decade of the *ancien régime* that it became commonplace for French universities to incorporate calculus within the natural philosophy curriculum, and the first professor to have been found to definitely teach it at all was Nicolas Beguin, in the early 1770s. See Brockliss, *French higher education*, p.385.

23. The level of Voltaire's mathematical expertise is a subject open to some debate. On one hand, it has been maintained that he regarded with some chagrin and envy the skills his friend Mme du Châtelet. See Esther Ehrman, *Mme du Châtelet: scientist, philosopher and feminist of the Enlightenment* (Leamington Spa, 1986), p.58. A more favourable presentation of his mathematical credentials has recently been offered however by Roger Pearson in *Voltaire almighty: a life in pursuit of freedom* (London, 2005). Regarding vulgarisations of Newton, Diderot himself had once planned an elucidation of the *Principia*, but abandoned the project when forestalled by Messrs LeSeur and Jacquier (see editor's note in *Mémoires sur différents sujets de mathématiques*, p.114).

24. Krakeur and Kruger, 'Mathematical writings', p.223.

25. Diana Guiragossian Carr, 'The Image of Diderot in the British periodical press from 1750-1800', in *Diderot: digression and dispersion*, ed. Jack Undank and Joseph Herbert, p.61.

Diderot's interest in mathematics carried over to his work on the *Encyclopédie*. Although D'Alembert was responsible for most of the articles involving mathematics, Diderot did contribute to some degree by writing on Varignon's solution to the problem of perspective, Pascal's arithmetic machine, political arithmetic, and the articles 'Déficient', 'Parfait' and 'Géometrié Souterraine'. For the most part, these pieces involve verbal description of mathematical concepts, rather than any original or technical mathematical work. It is perhaps for this reason that Jean Mayer claims that they are relatively insignificant.[26] Nevertheless, the range of mathematical themes covered by these articles indicates the breadth of Diderot's mathematical grasp. Though he may not have been an expert in any of these fields, the fact that Diderot was able to treat intelligently such a wide range of mathematical ideas says much about his ability and the extent of his interests.

The article on political arithmetic is of special relevance because it heralds Diderot's involvement with the field of probabilities. This branch of mathematics began to take shape during the seventeenth century, largely due to the efforts of Blaise Pascal (1623-1662) and Pierre de Fermat (1601-1675), but it was during the latter half of the eighteenth century that mathematicians began properly to realise its potential to form the basis for a veritable 'science of society'.[27] The field was to be a major concern of Diderot's throughout the latter part of his life, and it dominates the second of his mathematical collections, *Nouveaux Mémoires sur différents sujets de mathématiques* (1761). Unlike Diderot's earlier memoirs, this work went unpublished, although the reasons for this are unclear. It is certainly the case that after his imprisonment at Vincennes in 1748, publication had become the exception rather than the rule for Diderot.

The fact that the *Nouveaux Mémoires sur mathématiques* were kept private would seem to have nothing to do with their quality. Indeed, the third and last of the articles, *Sur l'inoculation*, has been lauded as one of the finest, most eloquent and convincing arguments in favour of inoculation written during the eighteenth century.[28] Taking issue with D'Alembert's

26. Mayer, *Diderot, homme de science*, p.80. Of course, in considering Mayer's assertion it should be borne in mind that the *Encyclopédie* was intended for laymen, and as such we would not expect to find highly detailed mathematical demonstrations.

27. On the development of probability theory and its application to practical affairs during the latter half of the eighteenth century, see Keith Baker, *Condorcet: from natural philosophy to social mathematics* (Chicago, IL, 1975).

28. Mayer, *Diderot, homme de science*, p.81. This is no small praise, especially considering that the issue of inoculation aroused a great deal of interest in the eighteenth century, and was discussed by some of the finest minds of the time. Notable in this respect was the famous Swiss mathematician Daniel Bernoulli (1700-1782), with whom Diderot concurred regarding the necessity to proceed with mass inoculation. For more on the role of probability theory in Bernoulli's work, see O. B. Sheynin, 'Daniel Bernoulli's work on probability', in *Studies in the history of statistics and probability*, ed. M. G. Kendall and R. L. Plackett (London, 1977), p.105-33.

profound doubts regarding the practice of inoculation, Diderot chastised his colleague for confounding matters of public interest with those of particular individuals. More importantly, at least in the context of the present study, Diderot's support for the practice of inoculation stemmed in large part from his willingness to rely upon the science of probabilities in coming to decisions about matters crucial to the public interest. In this, his attitude was markedly distinct from that of D'Alembert, who was tremendously suspicious of probabilities and refused to place matters pertaining to life and death in the hands of calculations that did not lead to certainty.[29] The basis of Diderot's challenge then is not inherently mathematical, but rather philosophical. The argument of *Sur l'inoculation* relies not so much on actual calculation but on particular notions regarding what intellectual tools are appropriate and effective in dealing with the issue at hand.[30]

The same cannot be said of the other two *mémoires* in the collection. In *Sur les probabilités* Diderot again inveighs against his mathematician friend D'Alembert. The problem in question surrounds the calculation of gambling odds, and as such the subject matter of this *mémoire* is not only more frivolous than that of *Sur l'inoculation*, but in many ways more theoretical as well. It is, however, no less interesting. Diderot takes D'Alembert to task for asserting that the chance of flipping a coin twice and not having it come up heads at least once is only two to one. The basis of D'Alembert's reasoning was that if the coin came up heads the first time then the game would be automatically resolved, and therefore there was only a 50 per cent chance that it would need to be decided by a second throw. Diderot took an altogether more sound tack, insisting that all possibilities (h-h, h-t, t-h, t-t) had to be considered before the first throw was made, and that each of these possibilities was to be regarded as equally likely, in a purely mathematical sense. In a calculation that has since been vindicated by modern algebra, Diderot demonstrated that the true odds that should be given for not throwing a head are three to one.[31]

Nouveaux Mémoires sur mathématiques also includes a piece entitled *Réflexions sur une difficulté proposée contre la manière dont les newtoniens expliquent la cohésion des corps, et les autres phénomènes qui s'y rapportent* (hereafter referred to as *Sur la cohésion*), a work that differs radically from the other two in terms of both subject matter and method. In this *mémoire*, Diderot again

29. Wilson, *Diderot*, p.431.

30. The issue of inoculation will be treated in a slightly different context later in this book (see chapter 5, p.187-88).

31. The essence of Diderot's calculation is that if the sum of the bets equals P, then the chances of Pierre (who is betting Jacques that a head will be thrown) are 3P/4, while those of Jacques are P/4, or odds of 3:1. Modern algebra would state the matter thus: the chances of no head being flipped in one trial is 1/2, in two trials it is (1/2) squared, therefore the chance of tossing at least one head in two throws is 1−(1/2) squared = 3/4. For a detailed discussion of this dispute, see Krakeur and Kruger, 'Mathematical writings', p.224.

demonstrates a profound awareness and ability to deal with issues surrounding Newton's *Principia*. Here Diderot comes to the defence of Newton, or more precisely to the defence of the Newtonian world-view. It is an argument against those who, unable to reconcile the law of universal gravitation with the force by which small bodies cohere, had proposed that Newton's inverse square law was actually subject to variation. The most notable mathematician to suggest such a revision was Alexis-Claude Clairaut (1713-1765), who proposed that a/x^2 (the inverse square law) be superseded by $a/x^2 + a/mx^4$. His initial motivation for positing this change stemmed from attempts to solve the problem of the seemingly anomalous transit of the moon's apogee. Difficulties encountered in terms of reconciling calculation with observation eventually led Clairaut – for a time at least – to suspect that there may be at least two laws of attraction (perhaps even two distinct species of attraction) at work in the universe. Such variability of Newton's simple formula could then be used, according to Clairaut, to account for a wide variety of phenomena, including the means by which small bodies manage to cohere while remaining distinct from and not collapsing into one another.[32] Diderot's stance regarding this move is unequivocal. Though too tentative to proclaim that he can explain everything about the means by which small bodies cohere, he uses in *Sur la cohésion* two geometric demonstrations to suggest that variation of Newtonian law such as that proposed by Clairaut is unacceptable, that it is not in accordance with phenomena, and that bodies must cohere to one another on the basis of the same mathematically described law that governs the motions of bodies at a distance.[33]

Throughout his later years, Diderot continued to occupy himself with various mathematical projects and problems. These fragments, accumulated over the course of approximately twenty years, are now assembled in the Fonds Vandeul, and appear in his *Œuvres complètes* under the title *Mathématiques*. As with Diderot's earlier mathematical exploits, these later endeavours represent a wide range of interesting work, and provide further testament as to his competence as mathematician. They include articles on celestial mechanics, the geometry of infinity, annuities, the

32. For more on the relevance of this issue, especially to chemistry, see chapter 3, p.100, 102 and 103).

33. *Sur la cohésion*, *OC*, vol.5, p.13-18. It is worth noting that in addition to raising the ire of Diderot, Clairaut was also involved in debate with D'Alembert, Euler and Buffon regarding this issue. Buffon's objections to Clairaut are of special interest, as they demonstrate much about his critical attitude towards mathematics (in which he prefigured much of what Diderot had to say on the subject). Buffon argued that Clariaut's conclusions 'are based only on calculation, [but] one can represent anything with a calculation, and accomplish nothing; if one is allowed to put one or several terms at the end of an expression of a law of physics, as with that of attraction, it is [...] arbitrary instead of representing reality'. See Jacques Roger, *Buffon: a life in natural history*, translated by Sarah Lucille Bonnefoi (Ithaca, NY, 1997), p.53-57.

duration of human life, an outline for a lottery project and plans for the construction of a deciphering machine.

It is also during these later years that Diderot reprised his early fascination with the problem of the quadrature of the circle. In a number of propositions dedicated to 'cyclométrie', he delves into complex geometry and analysis in an attempt to determine whether or not a solution to the ancient problem is actually possible. These propositions see Diderot deal once again with Newton, as he rather humbly suggests that one of the latter's demonstrations is suspect.[34] In contrast to the brashness of Diderot's early confrontation with the problem of squaring the circle, this later, much more mature work is characterised by scepticism and indeterminacy. Such an attitude is mirrored by an autobiographical passage in *Réfutation d'Helvétius*, in which Diderot describes his fascination with the problem and relates his attempts to come to terms with it. The passage may be somewhat awkward and unclear, but it does indicate that later in life Diderot experienced an increasing amount of doubt as to whether or not the problem was actually solvable.[35] The conclusion to his propositions regarding 'cyclométrie' is unambiguous. Although he still asserts that time is well spent examining the possibility or impossibility squaring the circle, Diderot is amazed and confounded by the multitude of angles from which the issue can be considered. He goes on to postulate that attacking the problem involves a great many sophisms in which we can never arrive at a real discovery, and in the end claims that he would not be surprised if the answer could never be found. Indeed Diderot finishes by stating that the weighty geometrical work he has presented should not be regarded as a demonstration of anything but falsity.[36]

Of course, Diderot's doubt was well placed. As is now known, there is no solution to the problem of squaring the circle. The fundamental reason for this is that 'pi' is an irrational number, meaning that the value which ultimately defines a circle can never be precisely translated into a number that will represent the rational sides of a square. Curiously Diderot himself speculated on the possibility that 'pi' is irrational, although definitive statement of the fact had to await the work of Legendre (1752-1833) in the nineteenth century.[37] How tempting it is to delve momentarily into the realm of 'what if?' history and consider how Diderot's stock as a mathematician may have been dramatically raised had he connected these two speculations and been the first to conclusively demonstrate that it is in fact impossible to square the circle!

Such speculation is however both fanciful and unhistorical. What matters most regarding Diderot's canon of mathematical work is that it

34. *Mathématiques*, *OC*, vol.11, p.679-80.
35. *Réfutation d'Helvétius*, p.594-95.
36. *Première proposition de cyclométrie*, *OC*, vol.11, p.680.
37. Krakeur and Kruger, 'Mathematical writings', p.223.

demonstrates his status as a reasonably competent mathematician. He had a solid grasp of geometry, algebra, probabilities and calculus. Moreover, this knowledge usually extended to the most up-to-date methods. For example, he was able to employ continued fractions, which was then a recently introduced technique.[38] As Coolidge as well as Krakeur and Kruger have observed, Diderot was an intelligent, capable and original mathematician, who produced some interesting and often novel work. While his memoirs do occasionally digress into verbal descriptions relating to various moral or philosophical issues, Diderot was, on the whole, a praiseworthy mathematician; on occasion even able to get the better of professionals such as D'Alembert.

That Diderot possessed a thorough working knowledge of various branches of mathematics means that his opinions regarding the meaning and nature of the discipline came from an informed perspective. This is significant, for to anyone other than that rarest of breeds, the historian of mathematics, Diderot's mathematical ideas are not as interesting or as important as his ideas about mathematics. Jean Mayer claimed that Diderot's interest in the discipline was more technical than philosophical,[39] but this is belied by the fact that questions regarding what mathematics is, how it functions, and the means by which it informs human knowledge were of the utmost importance to Diderot.

Traditionally, Diderot has been associated with a predominantly negative view of mathematics. He is often held up, along with Fontenelle and Buffon, as one of the eighteenth-century's greatest critics of the discipline, as a kind of mathematical anti-hero.[40] This reputation is, of course, not entirely undeserved. Despite his early love of the subject, Diderot did rail against the dominion of mathematics on behalf of a more concrete, experimental, qualitative approach to the study of nature. He is also well renowned for sharply criticising professional mathematicians, particularly regarding their methodology and underlying presumptions about the operation of reality.

The ultimate basis of this criticism lay in Diderot's belief that mathematics is unreal; that it is something that does not actually exist in nature.[41] The 'real world', according to Diderot, was not essentially

38. Krakeur and Kruger, 'Mathematical writings', p.228.

39. Mayer, *Diderot, homme de science* p.65.

40. Fontenelle (1657-1757) stressed that mathematics is a purely human construction, and that it does not truly accord with anything that is in nature itself. He claimed that 'the only things that exist in geometry are what we [i.e. humankind] put there, so to speak'. For more on Fontenelle see Roger Marchal, *Fontenelle à l'aube des Lumières* (Paris, 1997) and Aram Vartanian, *Science and humanism in the French Enlightenment* (Charlottesville, 1999). Buffon went even further, positing that mathematics is simply a tool without any inherent capacity to teach us anything about reality. On Buffon's anti-mathematical stance, see Roger, *Buffon*, p.37, 41, 52-58 and 86.

41. Of course, this may not have always been his view. In his first *Memoirés sur les différents sujets des mathématiques*, Diderot wonders whether or not there might be a 'natural

mathematical and any attempt to conceive of it as such he viewed as fundamentally misguided. Maintenance of this view placed Diderot in stark contrast to many of his contemporaries, some of whom followed the Pythagorean philosophy of asserting that numbers are real entities; that they exist in much the same sense as Platonic 'forms'. According to such a view, the precise relationships described by mathematicians were discoveries of relationships that actually existed in nature, or to put things another way, it was perceived that the 'book' of nature was written in the language of mathematics. An exemplar of this conception was Voltaire, who held that mathematics was innate to the human spirit, and that the work of the mathematician was that of mirroring reality, of capturing eternal truths that had objective existence outside of the human imagination.[42] An earlier and even more powerful exponent of this belief was Leibniz, who argued that mathematical ideas (although generated in the human mind) 'are relations that embody certain eternal truths, according to which the phenomena of nature are governed'.[43]

Along with Leibniz, the natural philosophers who left the greatest mark on eighteenth-century views regarding the mathematisation of reality were Descartes and Newton. The position of the former was very similar to that of Leibniz. For Descartes, geometry was a matter of more than just imaginative representation, it was seen as capable of capturing the underlying truths of reality.[44] Although the Newtonian position was marked by an admission that mathematical concepts are not necessarily real, but are an invention of the human mind, Newton and his followers nevertheless maintained that mathematics was the proper means by which to express and capture the relationships that existed between phenomena. In other words, despite any differences regarding the metaphysical status of mathematics, both Cartesians and Newtonians stressed the need to frame reality in mathematical terms; to reduce descriptions of nature and expressions of its laws to the language of mathematics. As James Barry has put it, their quest for a super-sensible reality achieved and expressed itself in the increasing primacy of a new

---

and secret trigonometry' which allows people to accurately estimate their distance from the point at which a particular sound originates. In *Les Aveugles*, he refers to 'L'Etre suprême, qui, selon l'expression ingénieuse d'un géomètre anglais [Joseph Raphson], *géométrise* perpétuellement dans l'univers'. It is essential to note, however, that Diderot immediately proceeds to stress that such simplicity, purity and perfection of understanding is not for human beings, and that the feebleness of our organs will never allow us to grasp things on such a level (*Les Aveugles*, *OC*, vol.2, p.179). For more on this aspect of Diderot's thought in the context of *Les Aveugles*, see Marian Hobson and Simon Harvey (ed.), *Lettre sur les aveugles: à l'usage de ceux qui voient; Lettre sur les sourds et muets, à l'usage de ceux qui entendent et qui parlent* (Paris, 2000).

42. Morin, 'Diderot et l'imagination', p.88.

43. C. J. Gerhardt (ed.), *Die philosophischen Schriften von Gottfried Wilhelm Leibniz*, 7 vols (Berlin: 1875-1890), vol.4, p.491-92.

44. Vincent Jullian, *Descartes, La Géométrie de 1637* (Paris, 1996), p.20 and 37.

mathematical realm, and 'the development of this new "mathematical space" is the very essence of the new science, for whatever is placed within it is automatically reworked and transformed'.[45]

This Cartesian/Newtonian/Leibnizian juggernaut was perhaps best represented in the mid to late eighteenth century by the likes of D'Alembert and Condorcet. Although more will be made in the next chapter of the discrepancies that existed between the philosophical outlooks of Diderot and D'Alembert, it suffices to say here that while D'Alembert acknowledged the abstract and constructed nature of mathematics (that is, he was not a Pythagorean), he placed great emphasis on the ability of this man-made system to serve as the key to understanding nature. In other words, for D'Alembert, mathematics amounted to more than just a highly qualified, strictly delimited tool, it was the preferred manner in which to express the truths of nature, or the true relations that existed between natural phenomena. The role of experience was downplayed in D'Alembert's scientific methodology: what mattered most to him was that reality be analysed, and that the results of this analysis be expressed in a manner as general, abstract and mathematical as possible. According to this conception, geometry and mechanics therefore emerged as more than just a part of science, but rather its essence. Arguably, this approach was taken even further by D'Alembert's protégé Condorcet, who was deeply committed to the view that mathematics was the pre-eminent means of arriving at the 'general truths of nature'.[46]

Diderot's stance, much like that of Buffon, seemed to approach the study of nature from a profoundly different angle to that expressed by men such as D'Alembert and Condorcet. For Diderot, emphasis remained on the issue of mathematics being an invention, a construct of human reason. As such, he perceived mathematical concepts as ideas that did not and could never correspond precisely with natural phenomena. Diderot maintained that mathematicians abstract away the qualities of individual objects, and that they conceive of things in such a general fashion that the items in question are stripped of everything that makes them unique. Diderot's canon is rife with comments about the purely intellectual status of mathematics, and about the fundamental separation that exists between mathematical concepts and reality. In *Suite de l'Apologie de l'abbé de Prades* (1752), for example, he states: 'je lui ferai considérer que s'il n'y pas un point, une ligne, une surface, un solide dans la nature, tels que la géométrie les suppose.'[47] Concurring with Buffon's *L'Histoire naturelle*, he

45. James Barry Jr, *Measures of science* (Evanston, IL, 1996), p.176.

46. For more on D'Alembert's application of and enthusiasm for mathematics see Thomas Hankins, *Jean D'Alembert: science and the Enlightenment* (Oxford, 1970), p.3, 24, 84 and 153. On Condorcet's attitude see Baker, *Condorcet*, p.107-109 and 117-18.

47. *Suite de l'Apologie de l'abbé de Prades*, *OC*, vol.2, p.635. This work was a defence of the unfortunate abbé de Prades, a theologian whose doctoral thesis was censured by the Sorbonne on account of its sensationalist and materialist implications. Significantly,

argues that mathematical abstractions only have reality in the mind of the mathematician who created them, and that the precision of the geometer can only ever approximate nature.[48] Diderot goes on to say that 'ces approximations sont communément d'autant plus éloignées du résultat de la nature, que les précisions ont été plus rigoureuses dans l'esprit du géomètre'.[49]

Without doubt, however, Diderot's most famous critique of mathematics is *Pensées sur l'interprétation de la nature* (1753-1754). It is this work, more than any other, which has earned him the reputation of being an opponent of mathematics. Almost from the start, *L'Interprétation de la nature* launches what seems to be a damning critique of the discipline, chastising its practitioners for putting too much stock in the veracity of their concepts and calculations, and not enough in the evidence presented by experience. A tone of apparent hostility is set very early:

> Une des vérités qui aient été annoncées de nos jours avec le plus de courage et de force, qu'un bon physicien ne perdra point de vue, et qui aura certainment les suites les plus advantageuses; c'est que la région des mathématiciens est un monde intellectuel, où ce que l'on prend pour les vérités rigoureuses perd absolument cet avantage quand on l'apporte sur notre terre. On en a conclu que c'était à la philosophie expérimentale à rectifer les calculs de la géométrie, et cette conséquence a été avouée même par les géomètres. Mais à quoi bon corriger le calcul de géométrique par l'expérience? N'est-il pas plus court de s'en tenir au résultat de celle-ci? D'où l'on voit que les mathématiques, transcendantes surtout, ne conduisent à rien de précis, sans expérience; que c'est une espèce de métaphysique général où les corps sont dépouillés de leurs qualités individuelles, et qu'il resterait au moins à faire un grand ouvrage qu'on pourrait appeler *l'Application de l'expérience à la géométrie*, ou *Traité de l'aberration des mesures*.[50]

This passage encapsulates perfectly the main thrust of Diderot's critique of mathematics: that there is a rift between the ideas of mathematicians and the objects of the real world, and those who ignore this and persist in trying to define the world according to the rigorous terms of mathematics will inevitably fail and fall into the trap of assembling an essentially false picture of nature.

It seems unsurprising therefore, that Diderot soon comes to a point at which he seems to question the very future of mathematics. *Pensée 4*,

Diderot himself is said to have had a hand in the preparation of the thesis, and thus to have been one of the prime instigators of this infamous affair.

48. As Jacques Roger argues, Buffon's *First Discourse* to the *Histoire naturelle* raised a question that we have still not satisfactorily resolved, namely, what is the precise relationship between mathematics and nature? Buffon perceived the relationship to be extremely problematic, viewing mathematics as a giant tautology that, however exact or sophisticated, is relatively useless in terms of its application to any of the 'true' sciences, such as physics (for Buffon, mathematics was a technique, not a science). See Roger, *Buffon*, p.86 and 91.

49. *Suite de l'Apologie de l'abbé de Prades*, p.635.

50. *Pensées sur l'interprétation de la nature*, *OC*, vol.2, p.716.

perhaps his most notorious piece of commentary on the discipline, is worth quoting at length:

> Nous touchons au moment d'une grande révolution dans les sciences. Au penchant que les esprits me paraissent avoir à la morale, aux belles-lettres, à l'histoire de la nature et à la physique expérimentale, j'oserais presque assurer qu'avant qu'il soit cent ans, on ne comptera pas trois grands géomètres en Europe. Cette science s'arrêtera tout court, où l'auront laissée les Bernoulli, les Euler, les Maupertuis, les Clairaut, les Fontaine et les D'Alembert. Ils auront posé les colonnes d'Hercule. On n'ira point au-delà. Leurs ouvrages subsisteront dans les siècles à venir, comme ces pyramides d'Egypte dont les masses chargées d'hiéroglyphes réveillent en nous une idée effrayante de la puissance, et des ressources des hommes qui les ont élevées.[51]

What emerges from *L'Interprétation de la nature* is a vision of natural philosophy whereby the work of mathematicians is equated to a game in which everything is determined by convention. Diderot claims that metaphysicians, for so long criticised by geometers for being men who know nothing ('Qu'est-ce qu'un métaphysicien? Un géomètre répondit: C'est un homme qui ne sait rien.') will have their revenge; chemists, physicists, naturalists and other practitioners of the 'experimental art' might easily apply the same definition to the work of geometers.[52]

Yet Diderot's critique is even more penetrating than this, and applies not just to mathematics but to the entire 'spirit of system'. Diderot is not simply attacking that part of the intellect which deals with mathematical concepts, but the entire method of arriving at knowledge through the process of deduction from abstract, fundamentally non-existent principles. It is, in other words, a scathing rebuttal of Cartesian rationalism. Demonstrating an extreme aversion to the construction of systems, Diderot repeatedly makes detrimental references to the 'philosophie rationnelle'. In *Pensée 23*, for example, he states that: 'la philosophie rationnelle pèse les possibilitiés, prononce et s'arrête tout court. Elle dit hardiment: *On ne peut décomposer la lumière*; la philosophie expérimentale l'écoute, et se tait devant elle pendant des siècles entiers; puis tout à coup elle montre le prisme, et dit: *la lumière se decompose*.[53] Elsewhere he claims that rational philosophers are unconcerned with facts, that they refrain from going down into the world and coming in contact with the objects which constitute reality, that they are perennially instructing (pontificating?) but never to any avail.

The anti-rationalist thrust of *L'Interprétation de la natur*e is nicely summed up in *Pensée 17*: 'Les sciences abstraites ont occupé trop longtemps et avec trop peu de fruit les meilleurs esprits; ou l'on n'a point étudié ce qu'il importait de savoir; ou l'on n'a mis ni choix, ni vues, ni

51. *Pensées sur l'interprétation de la nature*, p.717-18.
52. *Pensées sur l'interprétation de la nature*, p.717.
53. *Pensées sur l'interprétation de la nature*, p.728.

méthode dans ses études; les mots se sont multipliés sans fin, et la connaissance des choses est restée en arrière.'[54] Thus, assembly of knowledge on the basis of deduction from abstract principles, the pre-eminent example of which is geometry, bears the brunt of the polemic that is Diderot's *L'Interprétation de la nature*. His 'attack' on mathematics emerges as but the tip of an iceberg. It is the point of departure for demonstrating the vanity and misguidedness of presuming that the world conforms precisely to human ideas about it. This coincides very well of course with perceptions of Diderot as champion of qualitative, experimental natural philosophy.

It is equally important to note, however, that this image of Diderot as hero of an eighteenth-century anti-mathematical trend has earned him no small amount of criticism over the past two centuries. Many have branded the commentary of *L'Interprétation de la nature* short-sighted, pointing out that its assertions regarding the dim future of geometry would soon be mocked by the extraordinary advances of the nineteenth century (notably those of Gauss, who would open up new horizons in pure mathematics).[55] Diderot has also been criticised for attacking a paradigm of mathematics that was already waning, if not dead.[56] The argument here is that in condemning the 'geometric spirit' Diderot was actually at war with something more proper to the seventeenth century than the eighteenth. His contemporaries, it is said, were busy establishing the analytical method while he was left beating the 'dead horse' of pure mathematics.[57] The harshest line of interpretation then regarding Diderot's views on the subject would be that he was not only out of touch with the times, but boarding a rapidly sinking ship. Viewed as an opponent of mathematics, of quantitative science, and as ignorant of new methods of analysis, Diderot is often seen as representative of an approach to the study of nature that was on the path of inevitable decline, and was shortly to be totally eclipsed. He is typically cast as a romantic figure at odds with many aspects of modern science. Thus while Diderot may no longer be regarded as a mathematical philistine, the general impression conveyed by most commentators is that he is an unduly harsh critic of mathematics, in terms of both its method and its value, and that as a result of his own misunderstanding, misses the boat of progress.[58]

54. *Pensées sur l'interprétation de la nature*, p.726.

55. Mayer, *Diderot, homme de science*, p.101. For more on the achievement of Carl Friedrich Gauss (1777-1855) see W. K. Bühler, *Gauss: a biographical study* (New York, 1981).

56. Isabel Knight, *The Geometric spirit* (New Haven, 1968) p.19.

57. Jean Dhombres, 'Quelques Rencontres de Diderot avec les mathématiques', in *Colloque Internationale de Diderot*, ed. A.-M. Chouillet (1984), p.277. Ernst Cassirer also notes that the emphasis of mathematics had shifted in the eighteenth century from geometry to algebra (Cassirer, *The Philosophy of the Enlightenment*, translated by Fritz Koelln and James Pettegrove (Boston, 1955), p.54).

58. Jean Mayer, for example, characterises Diderot as moving in a direction opposite to that of the rise of 'modern' science.

These criticisms fail to take proper account of some very important points about Diderot and his relationship to mathematics. First, Diderot did have a sufficient grasp of what was going on around him. As we have seen from the survey of his work performed in the earlier part of this chapter, he was well versed in many contemporary mathematical methods, including those pertaining to algebra and even analysis. The most significant example of the latter is revealed by his interest in the science of probabilities. More importantly, although the tone of *L'Interprétation de la nature* may at times be rather severe, and although it may convey a predominantly negative impression of mathematics, it would be misguided to characterise his relationship to the discipline in such simple terms. While it is convenient to accept *L'Interprétation de la nature* as representative of his views on the discipline, and while the material contained therein is undoubtedly important, there is much more to Diderot's philosophy of mathematics – or better, another way of reading it. The requirement of taking a broader approach to Diderot's ideas on the subject is suggested, indeed, demanded by a number of problems arising from both his life and work. These problems confound the typical characterisation of Diderot as anti-mathematical, making it appear rather facile, if not wholly incorrect.

The first of these problems is that the intense love of mathematics experienced by Diderot during his earlier days continued throughout his life. In this, he was distinguished from Buffon, who virtually turned his back on the discipline after the 1750s.[59] While Diderot may not have pursued the subject with the same fervour in the 1770s that he had done thirty years previously, he pursued it nonetheless. Moreover, there is evidence that this pursuit was undertaken with no small degree of dedication. In 1775 Henri Meister, contributor to the *Correspondance littéraire*, wrote that 'Jamais M. Diderot n'a vécu plus solitaire que depuis son retour à Paris. Il s'est occupé tout l'hiver de calculs et d'algèbre. Ces études l'ont conduit dans ses heures de délassement à rêver aux moyens de composer une machine qui put servir d'imprimerie portative.'[60] Even stronger testimony regarding Diderot's continuing fascination with mathematics comes from his own *Réfutation d'Helvétius*. In an autobiographical passage, the same one in which he addresses his attempts to solve the problem of squaring the circle, Diderot speaks with great fondness of his mathematical endeavours.[61] In another section of the book he discusses the nature of genius and, tellingly, all of the geniuses that he cites as examples are mathematicians: Descartes, Leibniz, Newton, Galileo, Roemer.[62]

59. Roger, *Buffon*, p.58.
60. *OC*, vol.11, p.1085.
61. *Réfutation d'Helvétius*, p.594-95.
62. *Réfutation d'Helvétius*, p.561.

Even more difficult to reconcile with Diderot's apparent antipathy towards mathematics is the fact that he accorded the discipline a very prominent position within the tree of knowledge. This is best revealed by *Plan d'une université*. In this work, all branches of knowledge are causally linked, and mathematics comes first; it is the primary element of any education, the sphere of learning which provides the basis for all else. Diderot is unequivocal about the pre-eminent position of mathematical studies: 'Je commence l'enseignement par l'arithmétique, l'algèbre et la géométrie; parce que dans toutes les conditions de la vie, depuis la plus relevée jusqu'au dernier des arts mécaniques, on a besoin de ces connaissances. Tout se compte, tout se mesure. L'exercice de notre raison se réduit souvent à une régle de trois. Point d'objets plus généraux que le nombre et l'espace.'[63] In stark contrast to *L'Interprétation de la nature*, *Plan d'une université* presents the reader with a very positive appraisal of mathematics. Even geometry, which previously seemed to be the focal point of Diderot's critique of abstract rationalism in its isolation from reality, appears redeemed: 'Si l'on croit que la méthode des géomètres n'est pas applicable à tout, on se trompe.'[64]

Third, as Krakeur and Kruger suggest, mathematics (and geometry in particular) remained for Diderot the primary means of solution to physical problems.[65] The best evidence of this can be seen with reference to Diderot's own works, most notably *Sur la cohésion*. As we have seen, this piece makes extensive use of geometry in its attempt to suggest that Newton's inverse square law of gravitation need not be altered. Although Diderot lays out the philosophical basis of his argument in plain language at the beginning of the text, his reasoning is ultimately determined through geometric consideration of various of Newton's key propositions. At five pages in length, the elucidation of these geometric calculations comprises just over half of the short text. It is clear that, regarding the finer points of the behaviour of particles of matter from a purely mechanical, physical perspective, Diderot sees the use of geometry as invaluable.

All of these elements seem to mitigate the standard interpretation of *L'Interprétation de la nature*, and suggest that the manner in which mathematics fits into Diderot's world-view is much more complicated than a superficial reading of that work might suggest. But how is the aversion Diderot expresses towards the 'geometric spirit' in *L'Interprétation de la nature* to be reconciled with the rest of his life and work? One possibility is that *L'Interprétation de la nature* is anomalous, that it is a temporary backlash or rebellion against mathematics that cannot be taken as representative of his ideas about the subject. Supporting this theory is

63. *Plan d'une université*, p.771.
64. *Plan d'une université*, p.773.
65. Krakeur and Kruger, 'Mathematical writings', p.227.

the fact that the work may well be regarded as having been written 'off the cuff'. In the introduction, Diderot informs his readers that he is going to let words flow from his pen as the ideas flow from his mind.[66] If it is true that the work was so casually constructed, perhaps one should not demand of it a rigorous congruence with any of Diderot's other work? *L'Interprétation de la nature*, along with *L'Apologie sur l'abbé de Prades*, may just be a passing phase in Diderot's intellectual development, a random denunciation of mathematics on which he was to back-pedal in later years.

This is certainly how a number of scholars have read the issue. Jean Mayer, for instance, regards *L'Interprétation de la nature* as but one extreme of Diderot's thought, and argues that by the time *Plan d'une université* was written, Diderot had distanced himself a great deal from this critical position.[67] Such an interpretation is not, of course, implausible, but it is unsatisfying. Diderot expressed reservations about the absolute value of mathematics throughout his career, both before and well after *L'Interprétation de la nature*. Even *Plan d'une université*, which emphasises the importance and indispensability of mathematics, notes that 'Si l'on a souvent reproché au géomètre d'avoir l'esprit faux; c'est que tout entier à son étude, les choses de vie lui sont inconnues'.[68] Moreover, by simply arguing that Diderot's views regarding the nature of mathematics wavered throughout his career, attention is removed from what may be the most important aspect of the issue: the philosophical basis of both Diderot's negative and positive proclamations about mathematics.

The key to reconciling Diderot's mistrust of the 'geometric spirit', exemplified by *L'Interprétation de la nature* with all of the countervailing evidence attesting to his continued interest in the discipline, lies in his views regarding the nature of abstraction. As we have seen, Diderot regarded himself as something of an opponent of abstraction, but this opposition was not absolute. *Lettre sur les aveugles*, for instance, which deals in some depth with the formation of abstract notions, acknowledges the essentiality of the process to the acquisition of knowledge. Abstract ideas, whether relating to mathematics, to notions of God, or to morality, are dependent upon the senses of the thinking subject, and as such may be relativised, but Diderot is far from saying that this makes such ideas meaningless. Indeed the very fact that our abstract ideas are so affected by the type and manner of sensory data with which we are presented is the very thing that makes the way in which blind people construe the world so interesting to Diderot.

The fundamentality of abstraction to human thought and experience is also highlighted in the *Encyclopédie* article 'Encyclopédie' (written in 1755,

66. *Pensées sur l'interprétation de la nature*, p.715.
67. Mayer, *Diderot, homme de science*, p.101.
68. *Plan d'une université*, p.774.

around the same time as *Pensées sur l'interprétation de la nature*), in which Diderot makes the seminal point that the distinctions posited by human beings in order to render the universe comprehensible are arbitrary, but necessary.[69] Knowledge of the world cannot be formed without abstract thought. Conceptualisation is a process that underlies all human endeavour, be it artistic or scientific.

Similar views are expounded in the *Prospectus de l'Encyclopédie*. Here Diderot acknowledges that all human knowledge is ultimately subjective and that it is a matter of convention. There is a sense in which people do not discover relationships between the various phenomena which constitute reality so much as construct them. Taken together with the article 'Encyclopédie', the *Prospectus* argues that we live in a world in which we ourselves are the source of all meaning. In a statement which seems to have more affinity with twentieth-century existentialism than with eighteenth-century philosophy, Diderot notes that, were it not for the presence of Man, the universe would be completely silent and without inherent interest.[70] Man may be, to use one of Diderot's own phrases, 'un être qui passe comme l'éclair et qui rapporte tout à sa durée',[71] but this does not render the act of relation null and void. It is subjective, relative perhaps, but not meaningless, nor useless, and certainly not without interest to Man himself.

The development of concepts is therefore a matter of convention, but convention is the only means at the disposal of human beings for fixing the things of nature. This notion of fixing is important, for science is not just about the collection of facts but about assembling these facts into some kind of coherent picture. It is about imposing order (even if it is inevitably artificial and anthropocentric) upon the universe. This act of imposition, of making sense out of reality involves a movement in the mind from the particular to the general; from that which is ultimately identical only to itself to that which can be expressed by the use of signs.[72] Everything, Diderot tells us, the abstract and the concrete alike, is translated into signs.[73] It is only in this fashion that things can be related to one another and understood, much less communicated. Thus, according to Diderot, while abstract ideas may have no absolute value, while they may not deal with a priori concepts or essential truths, they do exist as a function of the human mind. They are the basis upon which humanity assembles a picture of the world. Diderot is therefore not

69. 'Encyclopédie', p.406.

70. 'Une considération surtout qu'il ne faut point perdre de vue, c'est que si l'on bannit l'homme ou l'être pensant et contemplateur de dessus la surface de la terre, ce spectacle pathétique et sublime de la nature n'est plus qu'un scène triste and muette' ('Encyclopédie', p.408).

71. *Sur deux memoirés de D'Alembert*, *OC*, vol.5, p.31.

72. 'Encyclopédie', p.416.

73. 'Encyclopédie', p.383.

against abstraction per se, but merely the consideration of reality in terms of purely abstract entities that are taken to precede experience. The intellectual process of abstract thought is therefore legitimised, as long as its nature and limits are kept in perspective.

So it is with mathematics. As Diderot described in his *Prospectus*, the science of nature (comprised of 'physique, mathématique et métaphysique générale') is constituted by the use of the senses to obtain knowledge of individual entities, the contemplation of which leads to the development of general ideas such as colour, sound, density, hardness, elasticity, gravity, movement, duration and quantity.[74] Mathematics is simply the sphere of human knowledge that deals with the last of these abstractions.[75] The concept of number and the rules governing the interrelation of numbers may not therefore be found in nature itself, but it is necessary for Man to induce these ideas from reality in order to be able to deal with it in a constructive manner. The discipline may be but an artificial product, but it is not without efficacy.

Viewed in this light, we should see *L'Interprétation de la nature* not as a sweeping condemnation of mathematics, but as an attempt to make a specific point about its nature and its use. Diderot may make his point forcefully, thereby creating a tone of hostility, but this should not detract from the more important issue: mathematics, a species of abstraction, is an important branch of the tree of human knowledge. *Pensée 2*, for instance, which emphasises that the region of mathematics is a 'monde intellectuel', does indicate that mathematics possesses some value. Reference to 'losing the advantage of mathematics' (if it is improperly regarded) infers that the discipline can indeed be of use, that there is some advantage to be had in its employment. Indeed, some of Diderot's comments in *L'Interprétation de la nature* may seem almost complimentary. When he states that geometry is near an end, he may not mean that the discipline will soon become useless, but that it will have reached a stage where, as an abstract system, it is complete. Thus, rather than being a relic of human thought no longer applicable to the study of nature, mathematics may be seen within the context of Diderot's thought as a fully developed tool, which, employed with proper circumspection, can contribute greatly to the task of making sense of the universe. Not being able to go beyond the 'pillars of Hercules' may not be such a bad thing.

Such a reappraisal can be applied to statements throughout Diderot's work. In *Nouveaux Mémoires sur mathématiques* Diderot claims that 'Toute la science mathématique est pleine de ces faussetés que M. D'Alembert

74. Once again, it might be observed that this idea is implicit in Diderot's thought from the time of *Les Aveugles*. The fact that, from Diderot's point of view, these notions have their ultimate origin in sensory evidence is the very reason that he argues that the abstract ideas of a person born blind will differ, at least slightly, from those of a fully sighted person.

75. *Prospectus*, *OC*, vol.2, p.306.

reproche à l'analyse des probabilités'.[76] This is not (as Krakeur and Kruger have portrayed it)[77] so much an attack on mathematics in general as a defence of probabilities. What Diderot is really attacking is the presumption of mathematicians who believe that their calculations actually embody reality, and that the rigorous nature of 'hard' disciplines such as geometry and algebra makes them innately superior to 'soft' disciplines such as probabilities. Diderot wants to claim that all mathematics is 'soft'. What is important to add, however, is that this is not cause to rule mathematics out as a potentially useful sphere of knowledge. To do so would be to throw the baby out with the proverbial bathwater. Diderot does not want to bring a halt to mathematical inquiry any more than he wants to stop playing with probabilities.

Thus, while what may be described as a crusade against the domination of mathematics in affairs of natural philosophy has come to be seen as one of the defining elements of Diderot's scientific career, the discipline was certainly not an aspect of human endeavour that he wished to rule out. Mathematics was not meant to be superseded by the 'qualitative sciences', but to coexist and indeed co-operate with them. Geometry, algebra, calculus and probabilities had a very definite place within Diderot's philosophy of science. What remains is to determine exactly what this place was, and to delineate what Diderot's works actually demonstrate about the role mathematics should play in human affairs and the world of science.

First, Diderot saw mathematics as possessing enormous pedagogical value. This is readily apparent from *Plan d'une université*, which, as we have seen, posits that arithmetic, algebra, probabilities and geometry provide the basis for all other learning. Without mathematics there is no foundation for rational mechanics (the second class in the first course of studies), which in turn leads to astronomy, then experimental physics and after that natural history, chemistry and anatomy.

Perhaps the most striking aspect of Diderot's plan is that logic and 'la grammaire générale raisonnée' is only the seventh element of the first course of studies.[78] This is not because he placed little value on the development of clear and rational thought, but because he saw mathematics as performing this role better than traditional logic instruction. Diderot perceived mathematics as the most powerful form of training for the mind; for him, it had an unparalleled capacity to sharpen the intellect. Those who could successfully manipulate the concepts of geometry and algebra were therefore equipped to deal with virtually any intellectual matter.

Despite this, mathematics was in Diderot's eyes eminently accessible

76. *Nouveaux Mémoires sur différents sujets de mathématiques*, p.31.
77. Krakeur and Kruger, 'Mathematical writings', p.230.
78. *Plan d'une université*, p.770.

and universal; it constituted not only the most profound but also the simplest and most basic form of reasoning: 'Je crois qu'il est plus aisé d'apprendre l'arithmétique et la géométrie élémentaires qu'à lire. Les lettres de l'alphabet ont fait verser aux enfants plus larmes, comme caractères de l'écriture, qu'elles ne leur en feront verser comme signes algébriques.'[79] It was therefore appropriate, according to Diderot, to begin teaching children the rudiments of mathematics as soon as possible. This, of course, brings to mind the manner in which he himself was subjected to arithmetic very early in life. He cites several examples of child prodigies, and argues that while mathematical geniuses such as these are not common, every child would be best served by an early grounding in the principles of mathematics. It is in light of this that he condemns contemporary curricula for lack of emphasis on mathematical training.[80]

Mathematics displaced for Diderot the Aristotelian syllogism and emerged within his system as the essence of logic, just as it had for Descartes a century earlier.[81] For both philosophers, it was undoubtedly the most perfect form of human thought. Although the clarity and unambiguousness of mathematics may have rendered its relationship to the material world of nature problematic for Diderot in a way that it did not for Descartes, Diderot did assent to the proposition that these qualities also endowed it with unparalleled strength. As *Plan d'une université* emphasises, 'La géométrie est la meilleure et la plus simple de toutes les logiques; la plus propre à donner de l'inflexibilité au jugement et à la raison.'[82]

This idealised perfection of geometry earns the discipline a hallowed place in Diderot's thought. The modern scholar might say that it becomes for him a paragon of enlightenment values in that its method is one which encapsulates and promotes reason. According to Diderot, familiarity with the geometric method illuminates the shadows of the mind and negates the follies of self-deception and confused thinking: 'Un peuple est-il ignorant et superstitieux? Apprenez aux enfants de la géométrie; et vous verrez avec le temps l'effet de cette science.'[83]

It is, in fact, in contrast to the rigorous and unforgiving logic of the geometric method that Diderot bemoans the haphazard state of language. As early as the article 'Encyclopédie', he observed that the state of human affairs would be much improved if all communication were as clear and logical as mathematics. Such a hope was illusory

79. *Plan d'une université*, p.771.

80. *Plan d'une université*, p.772.

81. On Descartes's distinction between mathematical logic and Aristotelian, syllogistic logic, see Jullian, *Descartes*, p.26.

82. *Plan d'une université*, p.773.

83. *Plan d'une université*, p.773. We might also consider this point in relation to the 'hieroglyphs' reference found in *L'Interprétation de la nature*.

perhaps (and Diderot acknowledges this), but it is worth noting that for Diderot mathematics was essentially nothing more than a type of language, an assembly of signs meant to connote certain concepts. In fact the notion of mathematics as stationary is given a genuinely positive connotation in the article 'Encyclopédie'. Diderot claims that as a 'dead' (that is, fixed or unchanging) system of signs, it is more perfect than any language.[84]

In addition to promoting the elevation of mathematics within the curriculum, Diderot also had some advice on how mathematics should be taught. In *Réfutation d'Helvétius* he lays out his plan:

> En sorte que cette élève possédait toute la géométrie élémentaire, persuadé qu'il ne savait qu'une seule proposition. En vérité, je préférais volontiers cette méthode à la méthode ordinaire. Toutes les vérités y sont rapportées vers un seul unique but qui leur sert de noyau. Ce noyau, c'est la massue d'Hercule, et les autres vérités en sont comme les clous: c'est un tout que rien ne peut romper [...] Il me semble que la science s'en établit d'une manière plus compacte et plus ferme dans l'entendement, qu'elle effraie moins le disciple, et que peut-être soulage la mémoire. Si cela vrai de la géométrie, cela le serait peut-être egalement de la méchanique, de l'astronomie, et des autres parties de la mathématique, qui se réduirait ainsi à la solution d'un assez petit nombre de problèmes.[85]

This plan of teaching geometry by way of application to a few core problems may be the only truly novel aspect of Diderot's ideas about mathematics and pedagogy. Of course, the notion that mathematical instruction is the best form of training for the mind was fairly commonplace in mid-eighteenth-century France.[86]

The same might be said of the second significant function of mathematics within Diderot's world-view: its application to the realm of practical affairs. Mathematics was championed by Diderot as indispensable to virtually all of the mechanical arts, and his work is rife with testimony as to its utility. In his *Correspondance*, for example, he discusses the application of geometry to the design of church domes, emphasising the necessity of referring to both experience and mathematics. The structural stability of domes is best served, Diderot tells his correspondent Sophie Volland, when their construction engages both the carpenter and the geometer.[87] Further testimony to Diderot's fascination with the everyday application of mathematics is provided by his own work on such diverse projects as the sundial, the compass, the clavecin, the deciphering machine and the 'imprimerie portative'. Not to be neglected, of course, is Diderot's work on the *Encyclopédie*. The technicality of that

84. 'Encyclopédie', p.387.

85. *Réfutation d'Helvétius*, p.598-99.

86. This idea started to take root in the seventeenth century, under the influence of Cartesianism. See Brockliss, *French higher education*, p.382.

87. *Correspondance*, Georges Roth (ed.), vol.4 (2 September 1762), p.126-27.

work, with its detailed plates meticulously depicting the construction and operation of multifarious machines, is itself a testament to mathematical regularity and precision, and of course to the many uses of geometry and algebra. Mathematics seems to emerge within Diderot's world-view as the engine of technology.

It was not just material technology, but social technology as well that inspired Diderot. This is borne out by his aforementioned work on probabilities. His refutation of D'Alembert's argument against inoculation is not, of course, a matter of purely intellectual import, it had grave and controversial social consequences. Likewise, he finds a tremendous use for probabilities, which he calls the 'physio-mathematical science of life' in all fields related to good governance and the assurance of human welfare. In *Plan d'une université* he states that 'cette partie de l'enseignement [le calcul élémentaire des probabilités] est d'un usage immense dans les affaires de la vie; qu'elle embrace et les choses les plus graves et les choses les plus frivoles; qu'elle s'étend à nos vues d'ambition, à nos projets de fortune et gloire, et à nos amusements'.[88]

Perhaps Diderot's most vigorous defence of the practicality of mathematics however is to be found in *Réfutation d'Helvétius*. In response to the contention of Helvétius that men concur as to the veracity of geometric demonstrations because they are indifferent to matters of truth and falsity, Diderot exclaims: 'Indifférents! Demandez cela à l'architecte, au peintre, au perspecteur, au commis des finances, à l'ingénieur, au méchanicien, au maçon, au constructeur de vaisseau, à l'opticien, à l'arpenteur, au géographe, à l'astronome, à presque toutes les classes de l'Académie des sciences.'[89] Diderot's firm rebuttal of Helvétius here demonstrates that while he may have regarded mathematics as, ultimately, a matter of convention, it was also much more than this. It was a system of relationships that had tremendous practical value. According to Diderot demonstrations could not be concocted at random and the world was not indifferent to the truth or falsity of calculations.

Here, of course, a potential difficulty arises for Diderot's philosophy of mathematics, namely that which modern philosophers might refer to as the problem of conventionalism. If, as Diderot maintained, mathematics is something unreal – if it is an invention of the human imagination rather than something that exists in nature and is discovered – then why

88. *Plan d'une université*, p.774. Diderot's observance of the more frivolous aspects of the calculation of probabilities is highly apt. Much seventeenth- and eighteenth-century work related to this branch of mathematics was conducted within the context of game theory, Diderot's dispute with D'Alembert over the game of 'Pierre and Jacques' being but one example. Other games which attracted the interest of mathematicians included that of 'Croix ou Pile', and Daniel Bernoulli's 'St Petersburg Paradox'. Of course, gambling itself was a not insignificant aspect of eighteenth-century life; it was, after all, through helping to rig a lottery that Voltaire made his fortune.

89. *Réfutation d'Helvétius*, p.538.

does it 'work'? Why is nature responsive to mathematical analysis, in the sense that one cannot be (as Diderot himself asserted) indifferent to the 'truth' or 'falsity' of calculations? Must not the concepts of mathematicians correspond to something real if buildings are to stand, cannons are to fire and machines are to operate as they do? Put another way, one might take the successful applicability of mathematics as the guarantee that the relationship between the concepts contained therein are real, that they do have some kind of objective existence.

Of course, as we have seen, Diderot was adamant that the world is not inherently mathematical. How then was he to account for the efficacy of the discipline in terms of manipulating reality? It might be said that he did not, at least not by way of explicitly acknowledging the problem and offering an equally explicit solution. Given this, a modern philosopher of science may be tempted to proclaim that Diderot's failure to address the question properly represents the biggest weakness of his philosophy of mathematics. To make such an argument, however, would be both anachronistic and futile. It would be anachronistic since to judge Diderot's thought on the basis of twenty-first-century philosophy necessarily obscures what he was trying to do; futile because even though Diderot did not tackle the problem on precisely the terms outlined here, there does exist in his work an implicit response to the entire issue, and it is this that should concern us presently.

The answer rests, in part, with Diderot's conviction that not only mathematics but all human knowledge is to one degree or another a matter of abstraction. It comes back to his proposition that the world in which we are dealing is the 'comprehensible world', which is distinct from the 'real world'. It is to take the view that mathematics 'works', even though it is invented, because the entire realm in which it operates is itself one of invention. The sphere of activity in which mathematics is applied (that is, with regard to matters of technology, delineating the relations between things, etc.) is conceptual. One might say that the 'intelligible' world becomes transparent to the human understanding (mathematics is but an aspect of this understanding) precisely because it is a product of the human mind.[90] The world could be seen as configured according to this human system of interrelated ideas. But adhering strictly to the position that mathematics works simply because all aspects of the world as we perceive it are matters of human invention (that all knowledge and the entire sphere within which knowledge operates is a function of human intelligence, or that the only world that exists (for practical purposes) is

90. In propounding such a view, Diderot is staking out a position not dissimilar to current 'embodied mind' theories of mathematics, which argue that the abstract concept of number (the foundation of mathematics) springs from the experience of counting discrete objects, and that humans thereby construct, rather than discover, mathematics. For more on this school of thought, see George Lakoff and Rafael E. Nez, *Where mathematics comes from: how the embodied mind brings mathematics into being* (New York, 2000).

the 'intelligible world') would imply that one need not posit or be bothered about the existence or nature of the 'external' world at all. In other words, in following this route, one winds up advocating a philosophical idealism similar to that of Berkeley. This was something Diderot was anxious to avoid. Although he admitted in *Lettre sur les aveugles* that the Berkelian philosophy was seductive and difficult to combat, Diderot did want to posit a science that referred (in some manner) to a material reality.[91] There had to be a dialogue between nature and Man's attempts to come to grips with it. The problem therefore remains: how can mathematics be both a matter of convention and something that is useful in the 'real' world?

The solution to this problem, like Diderot's assertion that everything that human beings call knowledge necessarily involves the process of conceptualisation, has in large part been implied by what has gone before. Mathematics, for Diderot, is abstracted from reality – it may be a convention, but this convention has its roots in what actually exists. The presumption is that while mathematics may not exist *in* nature, it is an invented response *to* nature. As such, the discipline is fundamentally similar to machines or other man-made objects, as well as other forms of knowledge. Diderot's insistence that mathematical expressions do not have existence independent of the human mind and that they should never be mistaken for being equivalent to something 'real' was not meant to rule out the possibility that these expressions work as generalisations of certain things that actually exist. In other words, according to Diderot's epistemology, mathematics is posited as the crudest type of generalisation that humans have invented. It is the set of ideas that is at once simplest and most removed from the infinite complexity of concrete reality. Yet the seminal point remains: however simplified, however distilled, however many degrees removed from that which really exists, mathematics is still ultimately derived from the world of things. Putting it in properly philosophical terms, mathematics is for Diderot an empirical generalisation, and as such connotes both some manner of connection to reality, as well as the active, creative role of the human mind.[92]

91. *Lettre sur les aveugles*, p.192.

92. Ernst Mach (1838-1916) employed the term 'empirical generalisations', regarding them as 'contingent truths which are confirmed by experimental evidence'. He further argued that 'these generalisations would be falsified if the results of experiments turn out to be different than hitherto observed' (see John Losee, *A Historical introduction to the philosophy of science*, 4th edn (Oxford, 2001), p.147). This notion might be related to Diderot's relatively simple proposition that mathematical calculations must necessarily be confirmed by experiment. An interesting implication of Mach's view, in terms of its application to mathematics, is that as a system it could, theoretically, be other than it is. Such an implication was explored by Henri Poincaré (1854-1912), who indeed argued that it is only a matter of convention that pure geometry is used to describe spatial relations among bodies. (See Losee, p.152., and also Henri Poincaré, *Science and hypothesis*, translated by G. B. Halsted (New York, 1905), p.39).

These issues come into even clearer focus upon examination of Diderot's views regarding the role played by mathematics in the interpretation of nature. This is the third significant function of the discipline within Diderot's world-view and, in the context of the present book, arguably the most interesting. That mathematics has a function relevant to the interpretation of nature is clearly demonstrated by Diderot in a number of works, through claims such as that in *Plan d'une université*: 'Point de mécanique sans géométrie; point de physique expérimentale sans quelque teinture de mécanique'.[93] In *Sur la cohésion* Diderot states: 'Je ne parle qu'en doutant; parce que dans une matière comme celle-ci, à moins qu'on ne soit géomètre très profond, il est très facile de se tromper.'[94]

In the *Prospectus* to the *Encyclopédie* Diderot divides mathematics into three primary categories: pure mathematics, mixed mathematics and physico-mathematics.[95] The first of these, as one might imagine, has no application to the study of nature, at least as far as Diderot is concerned. For him, that part of the discipline is totally cut adrift from reality, because it is nothing more than a series of abstractions referring only to themselves. It is a system of artificial regularity that makes no attempt to connect itself to the real world. Pure mathematics is concerned exclusively with what happens in the mind of the mathematician. Diderot's lifelong interest in the problem of squaring the circle fits neatly (no pun intended!) into this category.

Mixed mathematics is that realm which attempts to combine the abstract, self-referential concepts of pure maths with quantified elements of reality. It is a mixing of signs that refer to the abstract, non-existent entities of pure mathematics with those that refer to the concrete. The inevitable result, if we heed statements made by Diderot elsewhere in the *Prospectus* and in *L'Interprétation de la nature* ('En géométrie une quantité réelle jointe à une quantité imaginaire donne un tout imaginaire'),[96] are abstract signs. This is not to say, of course, that these signs are useless. Because there is an element of mixed mathematical abstraction that is derived from reality (certain of the numbers participating in the calculation are taken to represent real entities), the fruits of this discipline can be brought to bear upon certain, strictly delimited (i.e. quantifiable) aspects of the world. The prime example of what Diderot terms mixed mathematics would be the science of mechanics.[97] This science may not

93. *Plan d'une université*, p.780.

94. *Sur la cohésion*, p.12.

95. *Prospectus*, p.307.

96. *Pensées sur l'interprétation de la nature*, p.770.

97. In ascribing the label 'mixed mathematics' to this kind of work, Diderot was very much in line with contemporary usage. It should be observed however that the term 'mixed mathematics' was soon to be superseded by 'applied mathematics' – a phrase that undoubtedly makes much more sense to the twenty-first-century reader. For more on the meaning of 'mixed mathematics' in an eighteenth-century context, and the manner in

function for Diderot as a tool of discovery, but it can be used to invent, and if the fruits of the process are validated by or in accordance with physico-mathematics, then the inventions will work.

It is therefore to the final strain of mathematics, physico-mathematics, that we must now turn our attention. For Diderot, it is this aspect of the discipline which holds out the possibility of numbers and calculations, even geometry, acting as assistants in the task of rendering the universe comprehensible. Fundamentally mathematics serves the investigation of nature as a way of framing things, as a way of expressing the nature of relationships that exist between various phenomena. It is a model-making tool. There is a profound difference, Diderot tells us, between points on a piece of paper, and points on a piece of paper which actually represent real objects, for example the sun, the earth and the moon.[98] The former are only figments of the mathematician's imagination, they have no connection to or impact upon reality. The latter, however, constitute a model of reality.

It is important to emphasise here that the act of 'framing' natural phenomena in mathematical terms is actually descriptive; as has been witnessed with regard to the problem of conventionalism, mathematics is a credible and legitimate attempt to engage with reality by approximating it. For Diderot, the comprehensible world bears a direct relation to the real world in that the concepts of which it is constituted are induced from experience and then are subsequently used to describe that experience. Were mathematical expressions not turned back upon the world in this way, then the entire discipline would be subject to the same criticisms laid against pure mathematics. Thus the act of using mathematics to help make sense of the world is a two-stage process. It involves the derivation of the concept of number from experience, and then the turning of this concept back upon the raw, unordered chaos that is presented to us by our senses. This duality is clearly expressed in *Le Rêve de D'Alembert*:

Ce sont les signes du langage qui ont donné naissance aux sciences abstraites. Une qualité commune à plusieurs actions a engendré les mots vice et vertu. Une qualité commune à plusieurs êtres a engendré les mots laideur et beauté. On a dit un homme, un cheval, deux animaux; ensuite on a dit, un, deux, trois et toute la science des nombres a pris naissance. On n'a nulle idée d'un mot abstrait. On a remarqué dans tous les corps, trois dimensions, la longueur, la largeur, la profondeur; on s'est occupé de chacune des ces dimensions, et de là toutes les sciences mathématiques. Toute abstraction n'est qu'un signe vide d'idée. Toute science abstraite n'est qu'une combinaison de signes. On a exclu l'idée; en

---

which it became transformed into 'applied mathematics', see Gary Brown 'The Evolution of the term "mixed mathematics"', *Journal of the history of ideas*, (52:1) (Jan.-Mar. 1991), p.81-102.

98. *Réfutation d'Helvétius*, p.556-67. The reference here is, of course, to the work of Isaac Newton.

séparant le signe de l'objet physique; et ce n'est qu'en rattachant le signe à l'objet physique que la science redevient une science d'idées.[99]

That mathematics is part of this process, that it is an aspect of describing nature, is further emphasised by an earlier passage from the same work:

Diderot: Nous ne faisons qu'énoncer des phénomènes conjoints dont la liaison est ou nécessaire ou contingente; phénomènes qui nous sont connus par l'expérience; nécessaries en mathématiques, en physique et autres sciences rigoureuses; contingents en morale, en politique et autres sciences conjecturales.

D'Alembert: Est-ce que la liaison des phénomènes est moins nécessaire dans une cas que dans un autre?

Diderot: Non. Mais la cause subit trop de vicissitudes particulières qui nous échappent pour que nous puissions compter infailliblement sur l'effet qui s'ensuivra. La certitude que nous avons qu'un homme violent s'irritera d'une injure n'est pas la même que celle qu'un corps qui en frappe un plus petite le mettra en mouvement.[100]

Between these two passages a number of seminal points are encapsulated. The first makes it fully apparent that, as explained earlier, mathematics, for Diderot, has empirical foundations. Having ruled out the a priori existence of mathematics, Diderot demonstrates or at least suggests in general terms how it is induced a posteriori. Mathematics does not exist in nature but is derived from it; like every other concept or system of related concepts, its roots are to be found in sensory experience. Closely connected to this point is Diderot's insistence in the first passage that science cannot simply be about physical objects or signs alone, but must involve employing signs that are conjoined to and represent real things.

The latter of the two passages also emphasises the ultimate frailty of human knowledge which, as we have seen, is one of the guiding themes of Diderot's thoughts on the interpretation of nature and on mathematics in particular. We might think of the entire issue in terms of relative complexity. The universe, according to Diderot, is infinitely complex. Mathematics, on the other hand, is really nothing more than a way of simplifying this complexity. In the process certain subtleties will inevitably be lost, and with them any aspirations to attaining absolute 'Truth'. There are many different ways in which Diderot expressed this notion, but one of the most applicable here is his assertion that 'Un moyen presque sur de se tromper en métaphysique, c'est de ne pas simplifier assez les objets dont on s'occupe; et un secret infaillible pour arriver en physico-mathématique, a des résultats défectueux, c'est de les supposer moins composés qu'ils ne le sont'.[101] This, indeed, is a crux of Diderot's view of nature and of Man's attempt to understand it (there will always

99. *Le Rêve de D'Alembert*, *OC*, vol.8, p.147-48.
100. *Le Rêve de D'Alembert*, p.71-72.
101. *Lettre sur les aveugles*, p.179.

be 'inappreciable physical causes'). Man is ultimately consigned to representing reality in a manner that portrays it as being simpler than it actually is. The movement from the particular to the general which characterises Man's science comes down to the science of axioms, and the ultimate axiom is necessarily undefinable; it cannot be reduced any more, and it cannot really be proven.[102] Physico-mathematics, for Diderot, occupies an awkward middle ground between this ultimate undefinability and the raw, unordered, complex world of experience. This middle ground, however, is the best that Man can hope for. All human knowledge is, in a way, hypothetical.

Significantly, mathematics could be a key tool in the development and framing of scientific hypotheses. Diderot indicates in *Lettre sur les aveugles*:

> Mais que devons-nous penser des résultats du calcul? 1 Qu'il est quelquefois de la dernière difficulté de les obtenir; et qu'en vain un physicien serait trés heureux à imaginer les hypothèses les plus conformes à la nature, s'il ne savait les faire valoir par géométrie: aussi les plus grands physiciens, Galilée, Descartes, Newton, ont-ils été grands géomètres. 2 Que ces résultats sont plus ou moins certains, selon que les hypothèses dont on est parti sont plus ou moins compliquées. Lorsque le calcul est fondé sur une hypothèse simple, alors les conclusions acquièrent la force de démonstrations géométriques.[103]

Diderot goes on to assert that mathematical calculations can also be used to check hypotheses and to eliminate certain possibilities:

> Un avantage du calcul que je ne dois pas omettre, c'est d'exclure les hypothèses fausses, par la contrariété qui se trouve entre le résultat et le phénomène. Si un physicien se propose de trouver la courbe que suit un rayon de lumière en traversant l'atmosphère, il est obligé de prendre son parti sur la densité des couches de l'air, sur la loi de la réfraction, sur la nature et la figure des corpuscules lumineux, et peut-être sur d'autres éléments essentiels qu'il ne fait point entrer en compte, soit parce qu'il les néglige volontairement, parce qu'ils lui sont inconnus; il determine ensuite la courbe du rayon. Est-elle autre dans la nature que son calcul ne la donne? Ses suppositions sont incomplètes ou fausses. Le rayon prend-il la courbe déterminée? Il s'ensuit de deux choses l'une, ou que les suppositions se sont redressées, ou qu'elles sont exactes. Mais lequel des deux? Il l'ignore. Cependant voilà toute la certitude à laquelle il peut arriver.[104]

These passages demonstrate the possibility of fruitfully employing mathematical concepts as part of the investigation of nature, as well as how they must be regarded as part of an interplay with the real, physical objects that they represent.[105]

An even more startling testament regarding the applicability of mathematics to the interpretation of experience is Diderot's assertion: 'Il est

102. 'Encyclopédie', p.416.
103. *Lettre sur les aveugles*, p.190.
104. *Lettre sur les aveugles*, p.191.
105. *Lettre sur les aveugles*, p.190.

évident que la géométrie, en cas qu'il en fût instruit, lui fournirait un moyen infaillible de s'assurer si les témoignages des deux sens sont contradictoires ou non.'[106] Thus mathematical suppositions are kept in check by experience, but the relationship works both ways; mathematics can be used to confirm or reject the evidence provided by experience. It is a tool of judgement, and as such acquires some measure of authority in humankind's dealings with nature.

Scientific hypotheses and sensory experience were not the only domains in which mathematics had a role to play; the actual process of controlled experiment was also seen by Diderot as susceptible to the impact of mathematics. In an assertion paralleling that regarding the relationship between mathematics and (general, sensory) experience, Diderot noted that even though experiment may be required to temper the rigour of mathematical suppositions, experiments themselves are formulated and interpreted in an abstract, often mathematical fashion. Hypotheses, perhaps mathematically expressed, determine the nature and parameters of experiments to be performed. Results are measured and often only understood in mathematical terms.

It can now be appreciated how mathematics remained for Diderot, as Krakeur and Kruger have pointed out, the primary means of solution to physical problems. The use of mathematics to calculate, to frame and to describe the relationship of quantifiable natural phenomena to one another is the basis of *Sur la cohésion*. Even geometry, seemingly so maligned in other of Diderot's works, can be used to determine many aspects of how matter in motion behaves. *Réfutation d'Helvétius* also provides a number of good examples as to how Diderot perceived the operation of mathematics in the quest for knowledge of the universe. We search by experience and (often) express mathematically. Thus: 'Galilée voit tomber les corps; il s'aperçoit que leur vitesse s'accroît à chaque instant: il cherche par l'expérience quelle est la loi de cette accélération, et il découvre que les espaces parcourus dans des temps égaux sont comme la suite des nombres impairs'.[107]

Certainly Diderot perceived the role of mathematics in the interpretation of nature as being limited. Calculation could only be applied to matters dealing with quantity. Items that could be measured, such as size, distance, density and mass, were therefore susceptible to mathematical interpretation, to being described in a numerical fashion. Matters which dealt purely with quality, however, were not.[108] Also excluded of course

106. *Lettre sur les aveugles*, p.219.

107. *Réfutation d'Helvétius*, p.561.

108. *Prospectus*, p.306. An interesting question that arises here regards determining exactly where consideration of things in terms of quantity must end. What, exactly, can be measured, and what is insusceptible to measurement? Highly committed mechanists might try to argue that quantitative consideration can be driven very deeply into the heart of reality, suggesting even that organic matter is analogous to man-made machines and is

were inquiries into the causes of phenomena; mathematics was seen by Diderot as a discipline that could only be used to bring greater clarity to issues regarding effects. But for Diderot the key to ensuring that mathematics assists the development of human knowledge, rather than becoming a stumbling block, involved more than just adhering to strict parameters. Although, as we have seen, Diderot maintained that mathematically expressed hypotheses or models could never do any more than approximate nature, this does not mean that he accorded all such approximations the same value. While the mathematical physicist may never attain 'Truth', better correspondences with reality are always to be had. In a sense, Diderot perceived the relationship between the world and human knowledge of it as asymptotic; we may never arrive at our destination, but we can always get closer. To this end, Diderot advocated a process of constant resystematisation. In *Apologie de l'abbé de Prades* he asserts that the mathematician must be continually refining, reformulating, and adapting his model to the infinitely stubborn world.[109] Crucially, this continual resystematisation was not meant to apply only to the content of science but to its method as well.

A final element that enters into the reckoning then is ability, the ability of the philosopher employing mathematical concepts to use them carefully, dexterously and in an ever-increasingly refined fashion. Diderot placed emphasis on basic perceptual capacity as well – maintaining very forcefully in *Réfutation d'Helvétius* that the understanding of mathematical concepts is not equally clear to all. Some people, Diderot asserted, were simply more skilled than others in terms of perceiving relations, juggling abstract concepts and deriving from them proper conclusions. In *Lettre sur les aveugles* he argued that 'l'abstraction ne consiste qu'à séparer par la pensée les qualités sensibles des corps, ou les unes des autres, ou du corps même qui leur sert de base; et l'erreur naît de cette séparation mal faite, ou faite mal à propos'.[110] Thus, for Diderot, there are good systems and bad systems. What distinguishes the good from the bad is purely a matter of usage; abstraction well applied versus abstraction misapplied. Quite apart from Diderot's fairly strict criteria regarding cases of when mathematics can be used as a descriptive tool, the acceptability of mathematics in such matters hinges very much on considerations of the skilful manipulation of concepts in relation to the reality being described.

Thus, when in an article contrasting the approaches to nature of Diderot and D'Alembert, John Pappas upholds Diderot as a champion of 'the spirit of finesse', this should not exclude mathematisation.

---

therefore, in some sense, prone to a quantitatively based analysis. There is for Diderot, however, a question as to how far this is possible (at least on the basis of current knowledge and evidence).

109. *Suite de l'Apologie de l'abbé de Prades*, p.606.

110. *Lettre sur les aveugles*, p.179.

Although Pappas conceives of 'the spirit of finesse' in opposition to D'Alembert's 'esprit géométrique',[111] Diderot clearly felt that it is incumbent upon the natural philosopher to exercise finesse, skill and caution not only in framing experiments and interpreting results, but in applying mathematics to both of these processes. Presuming that mathematicians obey his injunctions, there is indeed a very crucial place for them in the act of interpreting nature.

For Diderot and his contemporaries, the most skilful and inventive manipulator of mathematical concepts was undoubtedly Isaac Newton, and it is with reference to Newton that Diderot's view of mathematics, its nature, limits, and applicability might be summarised. In his book *The Newtonian Revolution*, I. B. Cohen describes the various phases of what he calls the 'Newtonian style'.[112] The first of these phases involves exploring the mathematical consequences of a purely idealised system, the second relates the terms of this system to the data of observation or experimentally determined laws, and the third attempts to offer an explanation for how and why the system works in or as an aspect of the 'real world'. Cohen posits that part of the key to the success of Newton's method was his ability to distinguish properly between the first two of these phases; that is between the mathematical properties of a simplified construct that exists as nothing more than an analogy of nature, and the physical properties of the things accounted for in mathematical relations or rules or principles of the physical world as revealed by observation and experiment. It is further argued that subsequent to Newton many scholars and scientists blurred this crucial distinction, and it is with regard to this matter that Diderot enters the fray. One might see Diderot's body of work on the nature of mathematics as an attempt to reassert the Newtonian distinction between phases one and two and, more importantly, to downgrade the relevance of the first phase. Diderot wanted to retain a role for mathematics as one very useful form of expressing the relations between physical objects. The essence of his critique, even in *L'Interprétation de la nature*, was not to proscribe the use of mathematical concepts in human affairs and natural philosophy, but simply to circumscribe the legitimate aspirations of mathematical work, and to ensure that in addressing nature mathematicians did not attempt to raise their sights above and beyond the crucial conflation of physical object and mathematical sign.

That Diderot retained a place for mathematics within his natural philosophical framework, and that he thought long and hard about its meaning is significant, for his understanding of the discipline can now be seen as one of the cornerstones of his philosophy of nature. Consideration

111. John Pappas, 'L'Esprit de finesse contre l'esprit de géométrie: un debate entre Diderot et D'Alembert', *SVEC* 89 (1972), p.1229-53.

112. I. Bernard Cohen, *The Newtonian revolution* (Cambridge, 1980), p.55-63.

of Diderot's views on mathematics brings to the fore his ideas about the status of human knowledge. Many of Diderot's discussions about the abstract nature of mathematics bring into stark relief issues relevant to all human attempts at understanding the world. The *Prospectus* and the article 'Encyclopédie' make especially clear that the problems which beset mathematics also beset any other form of conceptualisation. Diderot believed that there was an absolute, metaphysical reality permanently beyond the comprehension of Man, but equally asserted that there is a sense in which Man can know the world. This sense was limited perhaps, and required the development of 'des signes plus analogues à l'étendue de notre esprit et à la conformation de nos organes',[113] but it was real and more importantly valid nonetheless. Diderot therefore held out the prospect that humanity is able (even if only from a finite perspective) to view the universe as noble, predictable and ordered. In so doing, he kept the way open for science. Without some epistemological middle ground between the boundless chaos of reality and the curiosity and ingenuity of humankind, there would be no chance for us to glean from the universe anything resembling a sensible picture. It is the attempt to ground mathematics and establish the parameters within which it must work that epitomises this struggle for Diderot better than anything else.

113. *Lettre sur les aveugles*, p.179.

2. Experimental physics

ALTHOUGH mathematics was, as the previous chapter has demonstrated, a discipline to which Diderot devoted a great deal of time and energy, a branch of science that has traditionally been associated far more closely with his natural philosophy is experimental physics. This is due in part to the fact that, as we have seen, Diderot's mathematical work and his thoughts about mathematics have traditionally been something of a niche interest. More importantly though, the comparative popularity of his ideas relating to experimental physics probably has much more to do with the fact that during the eighteenth century the term 'experimental physics' encompassed a great deal of subject matter. Put simply, almost any eighteenth-century thinker who purported to be interested in nature was bound to be somehow concerned with experimental physics. One need only to analyse the term itself, and all of the potential contemporary meanings of the words involved, to grasp how and why this would be the case. The label 'physique' had long been applied to a wide range of studies, and had equally subsumed a variety of methods. From the time of Aristotle, through the age of Newton and well into the eighteenth century, the word 'physick' or 'physique' had been broadly definable as the study of natural bodies, and as such encompassed everything from the study of matter in motion, to medicine, to what would later be understood as chemistry.[1]

It may be argued that in Diderot's day the term 'expérimentale' was hardly more clear. It did not, for example, necessarily refer to the kind of carefully controlled and monitored laboratory trials that twenty-first-century scientists normally associate with the word 'experiment'. For an early modern philosopher, the notion of experiment might have carried with it something of the modern sense, or at least something like it, though it may equally have connoted any kind of investigation (regardless of its precise method) that made use of instruments. More generally, 'experimental' or 'experimental method' might have been, depending on who you were talking to, anything based upon the evidence of the senses, or upon nature (however conceived), or upon observation of the natural world, or just the empirical method in general. Thus where the line might have been drawn between 'experiments' with a vacuum pump and 'experiments' in alchemy was not altogether clear. Indeed, for some it may not have even been clear that such lines really needed to be drawn.[2]

1. For more on the sense of scope implied by the term 'physics', particularly in an eighteenth century context, see Brockliss, *French higher education*, p.193.

2. Not only could the very definition of what it was to do an experiment be contended,

Given this room for conceptual manoeuvre, it has been postulated that for Diderot 'physique expérimentale' or 'philosophie expérimentale' (he used the terms somewhat interchangeably) was little more than a catchphrase incorporating a plethora of disciplines including (what today would be defined as) physics, chemistry and even biology.[3] Despite the breadth of meaning implied by the term 'experimental physics' however, it is necessary to comprehend precisely how Diderot perceived the discipline, the methods by which he thought it should proceed and the nature of its role in the interpretation of reality. Only once the parameters and substance of Diderot's interest in experimental physics are understood can it become apparent how this branch of natural philosophy fits into the framework by which he sought to grasp nature, and what consequences this held for his view of the world and its operation.

Such a task is made difficult not simply because of the broad definition usually accorded to the discipline of physics, but because the natural philosophical landscape of early mid-eighteenth-century France was in many ways remarkably fluid. Typically, the early part of the century has been cast as a battleground between the competing philosophies and world systems of Descartes (as propagated and embellished by his disciple Malebranche) and Newton, with, by the latter part of the century, the former being completely routed by the latter. The general lines of this story may be true, but as many scholars have convincingly demonstrated, the situation was far more complex than this.[4] First, the dividing lines between the Cartesian and Newtonian approaches to natural philosophy were not always clear, and an overly simplistic characterisation of either methodology is almost inevitably frustrated. Descartes, who is often taken as an exemplar of purely deductive, rationalistic thinking, was not entirely opposed to the practice of experiments, although the relative value of experiments were carefully qualified by a number of concerns and suspicions.[5] In many respects Newton, whose name was by mid-century often used as a rallying cry for the purposes of anyone who regarded themselves as practising natural philosophy properly, presents even greater difficulties. His *Principia mathematica*, the relevance of which was touched upon in the previous chapter, is an entirely different kind of work espousing a very different approach to the study of nature than his *Opticks*. Whereas the former is generally taken as the paragon of physics as

but the value of any given experiment or of its results could often be brought into question as well. An excellent discussion of the manner in which experiments were perceived in the early modern period is Steven Shapin, *Leviathan and the air-pump: Hobbes, Boyle, and the experimental life* (Princeton, NY, 1985).

3. Herbert Dieckmann. 'The First edition of Diderot's *Pensées sur l'interprétation de la nature*', in *Isis* vol.46, part 3, no.145 (September 1955), p.252.

4. See, for instance, Pierre Brunet, *L'Introduction des théories de Newton en France au XVIIIe siècle avant 1738* (Paris, 1931).

5. Desmond Clarke, *Descartes's philosophy of science* (Manchester, 1982), p.35-39.

a noble, mathematically driven, eminently cerebral exercise, the latter is often upheld as an example of a discipline that is gritty, tactile and heavily experimental, incorporating both the method of inductive reasoning at its very best, along with a healthy dose of intriguing speculation. Moreover, the process by which the Newtonian philosophy was assimilated in France (or, for that matter, anywhere else) was far from straightforward. This has been well illustrated by works such as Henry Guerlac's *Newton on the Continent*, which nuances the situation by demonstrating the slow, erratic, and often slightly confused process by which certain Continental philosophers adopted the basic parameters of the Newtonian world-view.[6]

This state of ambiguity is readily apparent upon investigation of the manner and context in which Diderot first became acquainted with the discipline of physics. Before proceeding to this investigation however, it should be noted that the scholar must carefully distinguish here between physics and experimental physics, for they were two very different things. The making of this distinction appears especially crucial in the case of Diderot, for his familiarity with experimental physics would seem to have come only much later and by means less formal than his exposure to general physics. This is not to say that it would have been impossible for him to acquire any understanding of the goals, methods and ramifications of experimental physics during his formative period – there were courses around – but simply that evidence does not permit the historian even to speculate about what Diderot might have known about them.[7] As things stand, there is nothing to suggest that experimental physics even entered into Diderot's consciousness until well after his university years. In considering the manner in which Diderot was introduced to matters involving the investigation and interpretation of the physical universe therefore, we must adopt a definition as broad as the subject itself (that is, as it was perceived at the time). That is to say that we begin by considering how and what Diderot learned of the general realm of physics during the initial stages of his intellectual development.

In addition to logic, ethics and metaphysics, the course in philosophy that Diderot followed at the Collège d'Harcourt in the early 1730s would have included at least a brief (one year) introduction to physics. In the 1730s this would have meant a predominantly Cartesian view of reality. Although in the early eighteenth century few, if any, Parisian professors

6. Henry Guerlac, *Newton on the Continent* (Ithaca, NY, 1981).

7. One of the most notable public courses on experimental physics to be held in Paris during the early years of the eighteenth century was that of the doctor, physicist and mathematician Pierre Polinière (1671-1734). Polinière is often considered to be the founder of experimental physics in France, and the popularity of his course (first conducted in 1709) is suggested by the fact that its published version, entitled *Expériences de physique*, ran into several editions over the course of the next few decades. See Pierre Polinière, *Expériences de physique* (Paris, Jean de Laulne *et al.*, 1741).

were wholly accepting of the Cartesian system of physics, not to mention methodology, discussion regarding the nature and behaviour of the physical universe would still have been conducted mainly in terms of Descartes's world-view. A mechanistic universe, notions of vortices and the plenum would have been at the centre of discussion. There also would have been little emphasis on mathematics, much less any attempt to describe phenomena in terms of mathematical formulae or mathematically expressed laws. While the ideas of Newton were beginning to penetrate the Académie des sciences by this time, they still had little or no impact whatsoever on the educational establishment. If the name of Newton presented itself, it tended to be brusquely pushed aside.[8]

Diderot's instructor in physics at this time seems likely to have been Pierre Lemonnier, who was a professor of philosophy at d'Harcourt from 1714-*c.*1750.[9] In light of Diderot's later intellectual development – which, as we shall see, carried him away from a predominantly Cartesian view of physics towards (and even beyond) a more Newtonian view – the prospect of this connection to Lemonnier may be significant. Lemonnier was by no means a Newtonian, at least not during Diderot's days as a student. He fervently denied the validity of the Newtonian picture, repudiated the concept of attraction and was not accepting of the Englishman's method. Despite this, Lemonnier was no friend of Descartes. He went further than most of his colleagues in terms of pointing out problems with the content of Cartesian physics, objecting even to some of its most fundamental elements. He rejected, for instance, Descartes's equation of matter and extension. He challenged the concept of the plenum and maintained the existence of 'philosophical vacuums'. Lemonnier also rejected Descartes's view of the elements, arguing that there could be no such thing as primary elements on account of matter's indefinite divisibility.[10] It may well be that Lemonnier planted for Diderot the first substantial seeds of doubt regarding the Cartesian system of the universe, which makes this course a potentially significant turning point in the career of one who was educated originally in accordance with the Jesuit *ratio studorium*.

Exactly how Diderot's education regarding matters of physics proceeded immediately after the Collège d'Harcourt is unknown. As we have seen in chapter 1, the details of his intellectual formation in general during the mid 1730s are extremely sketchy. By the 1740s however, the

8. See Brockliss, *French higher education*, p.185 and p.337-90.

9. Brockliss, *French higher education*, p.464.

10. Brockliss, *French higher education*, p.358-59. The basis for this argument (the indefinite divisibility of matter) would of course have been a factor which placed Lemonnier in opposition to Newton as well. By framing the argument as he did, Lemonnier was essentially using one aspect of the Cartesian world-view to counter another facet of that very same system.

picture presented is slightly different. By this time, Diderot would at last have begun to develop some awareness of experimental physics. Certainly he had become familiar (though perhaps yet not in detail) with a number of public lectures (and even perhaps private courses) given in Paris on the subject of experimental physics. Probably the most notable such course was that of the abbé Nollet (1700-1770). Alongside Voltaire and Maupertuis, Nollet was a key figure in the promotion of the Newtonian philosophy in France.[11] Indeed, Guerlac has claimed that his successful public lectures (first presented in 1735) were evidence of the 'final victory' of Newtonianism in that country.[12] These *Leçons de physique expérimentale* gave prominence to Newton's experiments on colour, but were probably most widely acclaimed for their spectacular and highly novel demonstrations pertaining to electricity, a relatively new field in which Nollet was regarded as the nation's leading expert. Given this element of fame, it is unsurprising that Diderot knew about Nollet and his course. What is much more difficult to determine is whether or not the *philosophe* actually attended any of the lectures, and if he did, what their content was. The only correspondence that exists between Diderot and Nollet revolves around the matter of certain plates used in the *Encyclopédie*, and provides no real information as to how Diderot evaluated the latter's theories and experiments. Perhaps the most relevant comment that can be made regarding Diderot's perception of Nollet is that the experimental physics section of *Plan d'une université* recommends Nollet's course as part of the curriculum.[13] From this, we might conclude that Diderot had some respect for Nollet as a natural philosopher, or at least as an experimenter.

The primary source of Diderot's knowledge of physics (both general and experimental) would seem to have come from personal reading. Although his library is now dispersed and it is impossible to identify all of the books that Diderot owned and may have read, Herbert Dieckmann and others have managed to gather some information on the contents of his library.[14] This list includes a number of significant works on the subject of physics: Voltaire's *Eléments de la philosophie de Newton* (1738), Claude Perrault's *Essais de physique* (1680-1688), Marc Thomin's *Traité d'optique, méchanique, etc.* (1749), Zimmerman's *Traité de l'expérience*, Maupertuis' *Venus physique* (1745) and the works of Francis Bacon comprise just a few. Significantly, Diderot was also a regular reader of the *Compte rendus* of the Académie des sciences, through which he would have

11. On Nollet see Jean Torlais, *L'abbé Nollet: un physicien au siècle des Lumières* (Argueil, 1987).

12. Guerlac, *Newton on the Continent* ,p.163.

13. *Plan d'une université*, p.781.

14. See Herbert Dieckmann, 'Observations sur les manuscrits de Diderot conservés en Russie', in *Diderot studies* 4 (1963), p.53-71; Arthur Wilson, 'Leningrad, 1957: Diderot and Voltaire gleanings', *French review* 31, p.351-56; Vladimir Sergeevich Liublinskii, 'Sur les traces des livres lus par Diderot', *Europe*, 405-406 (Jan-Feb 1963), p.276-90.

had access to much of the most innovative scientific thought of the day.[15] Finally, it must be noted that in addition to promoting the course of the abbé Nollet, Diderot's *Plan d'une université* prescribes for students the works of Pieter van Musschenbroek (1692-1761), Stephen Hales (1677-1761) and the *Opticks* of Newton.[16] On the basis of this, it is reasonable to assume that Diderot was familiar with the content of each of these works as well.

Something that must not be neglected is the fact that by the 1740s Diderot had, as we have seen, a good working knowledge of geometry, algebra and even calculus. Such a mathematical background equipped Diderot with the tools to comprehend even much of the more complicated physics of his day, including that of Newton's *Principia mathematica*. This, perhaps, provides the best clue as to the manner in which Diderot was originally exposed to and assimilated issues pertaining to the operation of the world on a physical level. Krakeur and Kruger suggest that it was mathematics which provided Diderot with a springboard to physics and to contemplation of issues pertaining to the nature of physical bodies.[17] What Krakeur and Kruger fail to do, however, is to explain how this was accomplished. In terms of Diderot's life, the point of transition between mathematics and physics is by no means clear. Indeed, the time at which his interests may have shifted substantially from the former discipline to the latter is no more apparent than the means by which he was originally exposed to either. One may trace the move from mathematics to physics in terms of Diderot's work, but exactly what this says is hard to determine.

The connection between mathematics and physics is, however, an important one, and as such it is valuable to survey the links between mathematics and physics within Diderot's schematic conception of the sciences. To do this, one must look at three key works: Diderot's *Prospectus* to the *Encyclopédie*, his article 'Encyclopédie' and his *Plan d'une université*. Each of these pieces attempts to lay out the relations between the sciences, establish their links and draw boundaries between them on the basis of the subject matter addressed and their manner of proceeding. By investigating Diderot's ideas as to how the sciences are theoretically connected, we may gain insight into the matter of how he himself moved from mathematics to physics.

Diderot's definition of 'physique' was intimately bound up with that of mathematics. In the *Prospectus*, he divides the 'science de la nature' into three primary branches: physics, mathematics and metaphysics.[18] To a

15. Jacques Proust, 'La Bibliothèque de Diderot', in *Revue des sciences humaines*, 33 (1958), p.260.

16. *Plan d'une université*, p.781.

17. Krakeur and Kruger, 'Mathematical writings', p.232.

18. *Prospectus*, p.306.

large extent, these categories are never entirely inseparable; at the very least, there are many points at which they are inextricably intertwined and lead naturally into one another. As Diderot explains, the basic operation of all three, indeed the basic operation of all human knowledge, is the process by which Man acquires knowledge of real bodies, which are individual and specific, and upon reflection develops ideas more abstract and general.[19] As Diderot states: 'Quant aux autres abstraites qui ne conviennent qu'à une certaine collection d'individus, ils ont constitué la science qu'on appelle physique.'[20] At root, the ultimate nature of physics is not really any different than that of mathematics; it is all about a movement in the mind from the particular to the general.[21] The primary difference is that mathematics involves moving farther away from reality, in that its concepts have become completely abstract and non-referential, in themselves, to any object in the 'real world'.

Physics is divided by Diderot, as for many of his contemporaries, into two sections: 'physique générale' and 'physique particulière'. General physics is the realm that deals with abstract ideas independently of the individual objects that have given rise to the abstraction in the first place.[22] It is here that physics bears its closest relationship to mathematics. Indeed, Diderot's distinction between general and particular physics is paralleled in mathematics by the division between pure mathematics, mixed mathematics and physico-mathematics. These categories do not line up exactly of course, as there is in physics no level of abstraction comparable to pure mathematics. Mixed mathematics and physico-mathematics however both involve the consideration of physical bodies in an abstract sense, that is, they deal with 'physique générale'. One might even say that mixed mathematics and physico-mathematics are tantamount to 'physique générale'. This notion is borne out as the passage on mathematics continues, describing in detail the subject matter of mixed and physico-mathematics; the two disciplines incorporate every sense in which physical bodies might be considered abstractly and independently.[23]

Particular physics is the science dealing with individual entities themselves, that is, as they actually exist. For this reason, it is placed outside of the bounds of mathematics. Diderot states in fact that 'physique particulière' must follow the distribution of natural history.[24] As such, each section of the science is understood as being 'De l'histoire, prise par les sens' of its subject. These subjects are extensive in number, ranging from

19. *Prospectus*, p.306. See also chapter 1.

20. *Prospectus*, p.306.

21. This process is of course far from straightforward, and will be discussed in greater detail shortly.

22. *Prospectus*, p.306.

23. *Prospectus*, p.306-308. See chapter 1, p.38-39.

24. *Prospectus*, p.308.

astronomy to geology, minerology, botany, zoology, medicine, physiology and even hygiene. It encompasses all aspects of the physical world, and as such accords very closely to traditional definitions of physics.[25]

The article 'Encyclopédie' does not alter this basic scheme. It reinforces the breadth of the discipline referred to as 'physique', claiming that it deals with an infinite number of problems.[26] It also makes clear that with the tools of mechanics and mathematics it attempts to seize upon the 'calcul vrai et réel des forces'.[27] This is essentially a repetition of Diderot's thoughts in the *Prospectus* regarding 'physique générale'; that an aspect of physics involves dealing with the idea of physical bodies, and attempts to comprehend their movements with the use of mathematics. It only remains, Diderot informs his reader, for experience to temper the rigour of these mathematical suppositions.[28]

The situation is presented under a slightly different guise twenty years later in *Plan d'une université* (1774-1775). This is appropriate, since in laying out the scheme for the *Encyclopédie* Diderot sought to establish the epistemological and methodological links between various disciplines, whereas *Plan d'une université* is driven by pedagogical considerations. It is nevertheless a useful indicator regarding Diderot's mature thoughts on the ordering of the sciences and it provides a simpler, clearer and perhaps more fundamental delineation of the role of each discipline within the framework of human knowledge.

The second element of Diderot's pedagogical plan (that is, the element immediately after mathematics) is the study of the 'Laws of movement. Falling bodies; free or on an incline. Centrifugal force and attraction. Mechanics and hydraulics'.[29] This is, in other words, rational mechanics; the application of mathematics to the most basic perceived movements of physical bodies. The third class listed in *Plan d'une université* extends this manner of inquiry to celestial phenomena. It is comprised of 'The Sphere and Globes. The system of the World. Calculation of Eclipses. Movement of Celestial Bodies, or astronomy. Gnomonsis etc.'[30]

It is not until the fourth class that Diderot comes to experimental physics, which he places here alongside natural history.[31] This recalls the proclamation of the *Prospectus* that the distribution of particular physics must follow along the same lines as that of natural history. Here, Diderot defines experimental physics as an imitation 'en petite' of the grand phenomena of nature. It is, in a way, a modelling process. It strives to illuminate the most general natural processes by investigating them on a

25. *Prospectus*, p.308.
26. 'Encyclopédie', p.448.
27. 'Encyclopédie', p.448.
28. 'Encyclopédie', p.448.
29. *Plan d'une université*, p.776.
30. *Plan d'une université*, p.777.
31. *Plan d'une université*, p.779.

small scale. In this way, experimental physics tests the principal agents of nature, which Diderot here lists as air, water, earth, fire, light, solids, fluids and movement.[32]

As was mentioned in the previous chapter, by way of demonstrating that mathematics has some role to play in physics, Diderot suggested in *Plan d'une université* that there is no mechanics without geometry, and no experimental physics without a tincture of mechanics.[33] There was clearly, in Diderot's eyes, a large degree of overlap between the disciplines of experimental physics and mechanics. Such a confluence is reiterated by Diderot's statement about the utility of experimental physics:

> La physique expérimentale s'introduit encore dans presque tous les ateliers des artistes. L'étude en est utile, agréable et facile. Point de machines, sans calcul de la solidité et de la fragilité; de la pesanteur et de la légèreté; de la mollesse et de la dureté; de la raideur et de la flexibilité; de l'humidité et de la sécheresse; du frottement et de l'élasticité. Les élèves verront les phénomènes, mais ils en ignoreront la raison, sans les connaissances préliminaires des deux premières classes.[34]

The final statement is significant, since the first two classes of the *Plan d'une université* are mathematics and mechanics. This confirms the fundamental relationship that existed for Diderot between mathematics, mechanics, and experimental physics.

It also, of course, complicates matters. Where does mechanics end and experimental physics begin? Also, what concerns, methods and preconceptions guide the experimental or even the general physicist in their work? For the answers to these questions, especially the latter, we must turn aside from the relatively neat presentation of the sciences as they are defined in the *Encyclopédie* or in *Plan d'une université*, and look at Diderot's seminal work of the subject of experimental physics: *Pensées sur l'interprétation de la nature*.

L'Interprétation de la nature was originally written in 1753, with a second, slightly revised edition composed and published in 1754.[35] It is laid out in a fashion reminiscent of Pascal's *Pensées* or of Diderot's earlier *Pensées philosophiques*; a collection of seemingly random and disconnected fragments. It will be recalled from chapter 1 that Diderot's preamble to the work claims that these fragments arose as the product of a spontaneous and casual process by which he let words flow from his pen as the thoughts presented themselves to his mind.[36] This has the air of a disclaimer, which would be appropriate, for as Diderot may have fore-

32. *Plan d'une université*, p.780.
33. *Plan d'une université*, p.780.
34. *Plan d'une université*, p.780.
35. For an excellent appraisal of the differences between these two editions, see Dieckmann, 'The First edition of Diderot's Pensées'.
36. *Pensées sur l'interprétation de la nature*, p.715.

seen, and as was certainly the case, *L'Interprétation de la nature* was harshly criticised by most contemporary reviewers on grounds of obscurity.[37]

The most striking aspect of *L'Interprétation de la nature*, or at least the one most commonly observed, is that it seems to draw a great deal of inspiration from Francis Bacon (1561-1626). Wilda Anderson, for instance, argues that Diderot self-consciously placed both the *Encyclopédie* and *L'Interprétation de la nature* under the aegis of the English philosopher.[38] It is undoubtedly true that at times the latter work appears to be something of a homage, and in terms of philosophical method it contains many elements that are consummately Baconian: a reverence for the practical, reliance upon the method of induction, as well as aversion to final causes and abstraction misapplied.[39] Most importantly however, *L'Interprétation de la nature* is pervaded by what may be termed a neo-Baconian empiricism.[40] Diderot argues on numerous occasions throughout the work that experience is the ultimate source of human knowledge, and indicates that those who rely on experience, collect facts and search for truth amongst that which is concrete and real are essential in best serving the needs and aspirations of Man. In *Pensée 20* Diderot states:

> Les faits, de quelque nature qu'ils soient, sont la véritable richesse du philosophe. Mais un des préjugés de la philosophie rationnelle, c'est que celui qui ne saura pas nombrer ses écus ne sera guère plus riche que celui qui n'aura qu'un écu. La philosophie rationnelle s'occupe malheureusement beaucoup plus à rapprocher et à lier les faits qu'elle possède, qu'à en recueillir de nouveaux.[41]

L'Interprétation de la nature continuously praises experimental philosophers for being those who make full use of their senses and who allow themselves to be guided primarily by experience. Diderot valorises the experimentalist by employing the metaphor of one who always proceeds according to the most basic instincts of touching, smelling and tasting. As

37. The only positive review was that of Diderot's friend Melchior Grimm, who heralded *L'Interprétation de la nature* as a work of genius.

38. Wilda Anderson, *Diderot's dream* (Baltimore, MD, 1990), p.12. Anderson also notes here that the very title of Diderot's work is reminiscent of three of Bacon's own, and that the epitaph ('Quae sunt in luce tuemur, E tenebris', from Lucretius' *De Rerum natura*) is a repetition of one by Bacon.

39. For Bacon's philosophy of science, see Antonio Pérez-Ramos, *Francis Bacon's idea of science and the maker's knowledge tradition* (Oxford, 1988), Markku Peltonen (ed.) *The Cambridge companion to Bacon* (Cambridge, 1996), and Thomas Kuhn, *The Essential tension: selected studies in scientific tradition and change* (Chicago, 1977), ch.3, 'Mathematical versus experimental traditions in the development of physical science', p.31-65. Not to be neglected is Bacon's own *Novum Organum* (the most recent English edition of which is edited by Lisa Jardine and Michael Silverthorne (Cambridge, 2000)), a tract that has been widely credited with helping to lay the foundations of experimental science.

40. See Lilo K. Luxembourg, *Francis Bacon and Denis Diderot: philosophers of science* (Copenhagen, 1967). Particularly useful is part 2, ch.2 ('Method and procedures', p.72-94) and ch.3 ('Concepts and interpretation of nature', p.95-106).

41. *Pensées sur l'interprétation de la nature*, p.727.

Pensée 10 argues, reason tends to consult itself, rather than nature, but 'L'instinct va sans cesse regardant, goûtant, touchant, écoutant; et il y aurait peut-être plus de physique expérimentale à apprendre en étudiant les animaux qu'en suivant les cours d'un professeur. Il n'y a point de charlatanerie dans leurs procédés.'[42]

Of course, Diderot's empiricism need not be seen as derived strictly from Bacon. His emphasis on the primacy of experience also betrays a connection to what were for him the comparatively recent sensationalist philosophies of Locke and Condillac.[43] The main point here is that for Diderot any consideration of human knowledge had to begin with the admission that everything is ultimately derived from experience. Innate ideas and a priori truths were ruled out; all knowledge had its source in information received through the senses. Thus, in Diderot's eyes, exploration of the world by means of the senses was not only the best way to proceed, it was the only way.

Of course, the collection of facts on its own comprised only part of the natural philosophical enterprise.[44] The indiscrimate collection of facts, unguided by any overarching considerations, would simply leave the natural philosopher buried beneath a mountain of data with no possible means of making sense of it. That Diderot was fully cognisant of this has been made apparent in chapter 1, where the role played by Man in the construction of knowledge as part of an attempt to bring some sense of order to the universe was discussed. Thus, although *L'Interprétation de la nature* does seem at times to come down firmly in favour of empiricism and in opposition to rationalism, there is no doubt that, for Diderot, some form of reason has a role to play in the interpretation of nature. While he frequently derided rational philosophers, portraying them as vain and impractical, Diderot was well aware of the need to deploy intellectual faculties such as logic, induction and the combination of ideas in the process of better understanding the world. In other words *L'Interprétation de la nature* presents the scientific pursuit as an activity that relies not just on sound empirical method, but careful use of rational methods as well.

This is nicely illustrated when Diderot carries his metaphor of the touching, smelling and tasting experimental philosopher a step further, and proclaims that while such a means of proceeding is eminently virtuous, the fact remains that the noble experimentalist is nonetheless blindfolded. It is, by contrast, the rational philosopher who sees, though he himself does not make use of his other senses:

42. *Pensées sur l'interprétation de la nature*, p.721.

43. On the influence of Locke in eighteenth-century France, the standard reference is John W. Yolton, *Locke and French materialism* (Oxford, 1991). For more on Condillac's sensationalism, see Knight, *The Geometric spirit*.

44. This was acknowledged of course by Bacon himself. See Michel Malherbe, 'Bacon's method of science', in *The Cambridge companion to Bacon*.

Nous avons distingué deux sortes de philosophie, l'expérimentale et la rationnelle. L'une a les yeux bandés, marche toujours en tâtonnant, saisit tout ce qui lui tombe sous les mains et rencontre à la fin des choses précieuses. L'autre recueille ces matières précieuses, et tâche de s'en former un flambeau: mais ce flambeau prétendu lui a jusqu'à présent moins servi que le tâtonnement à sa rivale; et cela devait être. L'expérience multiplie ses mouvements à l'infini; elle est sans cesse en action; elle met à chercher des phénomèmes, tout le temps que la raison emploie à chercher des analogies. La philosophie expérimentale ne sait ni ce qui lui viendra ni ce qui ne lui viendra pas de son travail; mais elle travaille sans relâche. Au contraire la philosophie rationnelle pèse les possibilités, prononce et s'arrête tout court.[45]

This was the state of affairs as Diderot perceived it in mid-eighteenth-century France, but it was not the state of affairs he wished to implement. Diderot's prescription was not to dispense with rational philosophy, but only to ensure that, after weighing possibilities and making pronouncements, it did not come to a full stop.

The metaphor of the blindfolded explorer, touching and feeling his way through the world is given even greater depth if considered in terms of Diderot's *Lettre sur les aveugles*, which deals with the faculties of one who is deprived of sight. Molyneux's problem, discussed toward the end of the work, even raises the issue of how touch and sight provide the mind with fundamentally different perceptions of reality and its manifold forms.[46] Diderot's inclination to side with Molyneux (and Locke) against Condillac in asserting that a person born blind who has recently been granted the power of sight will not be able to visually distinguish between a sphere and a cube underlines his insistence on the need to unite divergent faculties in order to acquire true and proper knowledge. It is in this capacity, among others, that reason becomes indispensable.[47]

Thus the collation and consideration of facts collected by experience is a requirement of any examination of reality, and this is made clear by Diderot on many occasions throughout *L'Interprétation de la nature*. Perhaps the best example is *Pensée 9*:

Les hommes en sont à peine à sentir combien les lois de l'investigation de la vérité sont sévères, et combien le nombre de nos moyens est borné. Tout se réduit à revenir des sens à la réflexion, et de la réflexion aux sens: rentrer en soi et en sortir sans cesse. C'est le travail de l'abeille. On a battu bien du terrain en vain, si on ne rentre pas dans la ruche chargée de cire. On a fait biens des amas de cire inutile, si on ne sait pas en former des rayons.[48]

45. *Pensées sur l'interprétation de la nature*, p.728.

46. *Lettre sur les aveugles*, p.215.

47. For more on Condillac's role as an importer of Lockean sensationalism to France, see Israel *Radical Enlightenment*, p.516-17, and Knight, *The Geometric spirit*. On Condillac's approach to the Molyneux problem see Knight, p.100.

48. *Pensées sur l'interprétation de la nature*, p.721. This passage bears a direct relation to Bacon's 'Bee analogy', found in bk 1, Aphorism 95 of the *Novum Organum*: 'Those who have treated of the sciences have been either empirics or dogmatical. The former like ants only

One of the primary themes to emerge from *L'Interprétation de la nature* then is the need to achieve a synthesis between the collection of facts and the application of reason to them. This is announced in the very first of the *Pensées* when Diderot declares that there are two types of philosophers: those with many instruments but few ideas, and those with many ideas but no instruments.[49] He goes on to state that the interests of truth demand that these two types be united, that the interpretation of nature can only proceed by drawing upon both. Ultimately Diderot is not attempting to favour one over the other. *L'Interprétation de la nature* does not seek for rational philosophy to be entirely supplanted by the experimental method, but for the two to become united. It is a work of reconciliation, aiming to harmonise thought, which is necessarily abstract, and the concreteness of experience.

The synthesis between rationalism and empiricism for which Diderot is arguing in *L'Interprétation de la nature* is even reflected in the title of the work. By referring to interpretation of nature (rather than investigation) it is made clear that the task of the natural philosopher is not one of simply uncovering facts, but of assembling them into a coherent picture. The scientist does not merely discover but creates, assembles, portrays. The act of understanding nature is the act of rendering it into comprehensible, human terms. It is, in a very literal sense, an act of interpretation. If a disproportionate amount of Diderot's condemnation falls on rational philosophy, this is perhaps only by way of redressing the balance between two previously competing means of interpreting nature. Rationalism bears the brunt of Diderot's criticism purely because it had so long been in the ascendancy; it needed, in Diderot's eyes, to be brought down, while experimental physics (then still the underdog, at least in France), needed championing.

This aspect of *L'Interprétation de la nature* is not particularly remarkable. Diderot's claim that the natural philosopher must embrace both empiricism and rationalism in due measure is tantamount to saying that the construction of knowledge is a process consisting of (at least) two stages: that of gathering information and that of appraising or combining it. Far from being revelatory, this is standard Lockean philosophy; the acknowledgement of both simple and complex ideas. Moreover Diderot's advocacy of a synthesis between the rational and experimental aspects of natural philosophy in *L'Interprétation de la nature* is not exclusive amongst his own body of work. Both before and well after 1753 Diderot made reference to the necessity of employing experiment and theorisation in the

heap up and use their store, the latter like spiders spin out their own webs. The bee, a mean between both, extracts matter from the flowers of the garden and the field, but works and fashions it by her own efforts. The true labour of philosophy resembles hers [...]'. (Francis Bacon, *Novum Organum*, in *Great books of the Western world*, ed. Robert Hutchins (Chicago, IL, 1952), vol.30, p.126.

49. *Pensées sur l'interprétation de la nature*, p.715.

construction of knowledge. *Essai sur les règnes de Claude et de Neron* (1782) contains a brief but illuminating passage on natural philosophical method in which Diderot states that the experimental philosopher collects facts in order to form a solid basis from which to speculate.[50] While Diderot presents this discussion partly as a contrast between experimental and rational philosophy, much as he does in *L'Interprétation de la nature*, it also makes explicit the need of employing both in the construction of knowledge. This is emphasised also in the article 'Encyclopédie' where Diderot states 'l'observation et la physique expérimentale multipliant sans cesse les phénomènes et les faits, et la philosophie rationnelle les comparant entre eux et les combinant, étendent ou resserrent sans cesse les limites de nos connaissances'.[51]

Where *L'Interprétation de la nature* emerges as more original and instructive is with regard to the manner in which Diderot advocates using sensory and intellectual faculties to assemble ideas. More than simply arguing that the scientist must employ methods both empirical and rational, Diderot presents a very clear concept as to how this is done. He outlines a platform for arriving at a knowledge of the physical world that rests on three pillars: observation, reflection and experiment. The basic relation between these three aspects of interpreting the natural world is laid out in *Pensée 15*:

> Nous avons trois moyens principaux; l'observation de la nature, la réflexion et l'expérience. L'observation recueille les faits, la réflexion les combine, l'expérience vérifie le resultat de la combinaison. Il faut que l'observation de la nature soit assidue, que la réflexion soit profonde, et que l'expérience soit exacte. On voit rarement ces moyens réunis. Aussi les génies créateurs ne sont-ils pas communs.[52]

The first of these elements, observation, would seem to need little in the way of elaboration. But even for Diderot (materialist and sensationalist that he was) the concept of observation was not necessarily straightforward nor free from need of explication. At least since the time of Descartes's *Meditations*, if not before, European philosophy had seen even the apparently mundane activity of gathering information by means of the senses cast into doubt. This is reflected in *L'Interprétation de la nature*, as Diderot speaks on several occasions of the feebleness of Man's organs.[53]

Moreover, as Diderot reminds his readers in *L'Interprétation de la nature* and elsewhere, not all observations are equal. Just as one theory may achieve closer correspondence with nature than another, so there is variance in the quality of observations. Some are accurate, some are not. The interpreter of nature must be careful, precise and assiduous even when engaged in the seemingly innocuous task of collecting 'facts'. Again,

50. *Claude et Neron*, *OC*, vol.13, p.588.
51. 'Encyclopédie', p.376.
52. *Pensées sur l'interprétation de la nature*, p.725.
53. *Pensées sur l'interprétation de la nature*, p.719.

one might see *Lettre sur les aveugles* as supporting *L'Interprétation de la nature.* It provides a good example of the complications which can plague the basic task of sensory acquisition. Raising questions regarding the means by which information is gathered, *Les Aveugles* makes it clear that even the process of observation cannot be taken for granted; it must be approached with due circumspection.

Thus it is that observations cannot always be taken at face value. Appearances can be deceptive, and the naive observer can easily be led astray. Ultimately observations require testing by means of experiment. The facts noted from experience must be manipulated and probed. Moreover, observations only truly acquire clarity and significance when placed in context by being brought into relation with other things. The notion of the lone 'fact' is anathema to Diderot's conception of natural philosophy; observations must be connected to one another, to experiments and to reflection in order to achieve meaning, place and indeed confirmation. Therefore the validity of observations at once underlies the construction of knowledge and is dependant upon it.

If the process of observation is complex, that of experimentation is even more so. The purpose of experiments, according to Diderot, is to confirm or reject suppositions formed on the basis of observations. In doing so, they play a crucial role in giving ideas solidity. As Diderot states in *Pensée 33*, with regard to some of his conjectures about magnetism, electricity and the nature of the globe:

> On peut avouer ou contredire ces notions, parce qu'elles n'ont encore de réalité que dans mon entendement. C'est aux expériences à leur donner plus de solidité, et c'est au physicien à en imaginer qui séparent les phénomènes, ou qui achèvent de les identifier.[54]

In some respects, experiments are at the opposite end of a chain from observation; they are a contrived interrogation of nature, where observation is more passive. Nevertheless, like observations, experiments are a very necessary part of what may be termed the empirical side of scientific work. The confirmation of ideas by means of experiment is just as crucial to the establishment of those ideas as the observations in which they are rooted. Again, the metaphor of experiments lending concepts solidity, weight or substance comes to the fore:

> Tant que les choses ne sont que dans notre entendement, ce sont nos opinions; ce sont des notions qui peuvent être vraies ou fausses, accordées ou contredites. Elles ne prennent de la consistance qu'en se liant aux êtres extérieurs. Cette liaison se fait ou par une chaîne ininterrompue d'expériences, ou par une chaîne ininterrompue de raisonnements qui tient d'un bout à l'observation, et de l'autre à l'expérience; ou par une chaine d'expériences dispersées d'espace entre des raisonnements, comme des poids sur la longueur d'un fil suspendu par ses deux

54. *Pensées sur l'interprétation de la nature*, p.736-37.

extrémités. Sans ces poids, le fil deviendrait le jouet de la moindre agitation qui se ferait dans l'air.[55]

Part of the problem with experiments is deciding exactly which ones to perform and, just as importantly, how they are to be conducted. Diderot posits that the veracity of any knowledge acquired regarding nature has much to do with the manner in which Man chooses to interrogate it. Just as observations vary in terms of quality, so experiments can be either well or poorly construed. This is a difficult and complex process, although Diderot does lay down some general guidelines for both the proper conduct of experiments, and for framing them intelligently.

First, and perhaps most obviously, Diderot is at pains to emphasise the importance of exercising patience, care and precision in the conduct of experiments. As we have seen, little is known about exactly what experience Diderot had of laboratory environments, at least at the time that *L'Interprétation de la nature* was penned.[56] It may well be that either through friends, contacts within the scientific community (such as Nollet) or through his work on the *Encyclopédie* he had by 1753 acquired some knowledge of the day-to-day operations carried out by experimentalists. Nevertheless, this remains a matter of presumption. In any case, *L'Interprétation de la nature* conveys the impression that much of the experimental work of the day was in need of great refinement.[57]

Part of this process of refinement was technical – having to do with the calibre of instruments used and the manner in which they were employed. The importance of instruments is emphasised in the first paragraph of *L'Interprétation de la nature*, which refers to rational philosophers as those with ideas and experimental philosophers as those with instruments. That Diderot saw the improvement of instruments as being desirable is witnessed on a number of occasions throughout the work. Certainly he regarded the inadequacy and imperfection of these technical aids to experiment as being amongst the primary obstacles to the expansion of Man's knowledge of the world.[58]

55. *Pensées sur l'interprétation de la nature*, p.720.

56. As shall demonstrated in the next chapter, Diderot most certainly did acquire some measure of familiarity with the day-to-day workings of a mid-eighteenth-century laboratory, but this can be confirmed only for the period post-dating his exposure to chemistry (an exposure that took place almost immediately subsequent to the composition of *L'Interprétation de la nature*).

57. Diderot was not alone in maintaining that much remained to be done in terms of improving experimental technique; most experimental philosophers of the mid eighteenth century would have acknowledged that the cognitive and technical aspects of their work could always be improved. Thus it was that throughout the century there existed debate regarding experimental styles, methods, standards and practice. For more on the development of the practical aspects of experimentation see Christian Licoppe, *La formation de la practique scientific: le discours de l'expérience en France et en Angleterre (1630-1820)* (Paris, 1996), especially ch.3 and 5.

58. *Pensées sur l'interprétation de la nature*, p.7 and 726.

The need to perfect instruments and to ensure the precision of measurements is emphasised most strongly in *Pensée 52*. Here Diderot characterises the apparatus used by natural philosophers as supplements to the senses, and argues that they are no less crucial to the acquisition of philosophical knowledge than the latter. He proposes that just as time should be spent examining the nature of sensory evidence, so it is important for the experimental philosopher to understand and to perfect his instruments.[59] Care and precision, both in measurement and the general conduct of experiments are the watchwords here; without these things the entire process of experimental philosophy is severely undermined.

But, as important and desirable as technical perfection is, Diderot's main concern in *L'Interprétation de la nature* is with the intellectual and epistemological aspects of the experimental process. The work includes a variety of suggestions designed to guide the experimental philosopher in his investigations. Perhaps the most important dictum is to try everything, to leave no stone unturned. According to Diderot, no investigation of nature is entirely fruitless, therefore, even if we only suspect that an avenue of inquiry may reveal something of value, it should be pursued. If we are failing comprehensively to uncover anything useful Diderot proposes trying the exact opposite of what we have been doing. Drawing an analogy between the rational and experimental philosophies to illustrate how this process of inversion can be helpful, he states that 'en philosophie purement rationnelle, la vérité est assez souvent l'extrême opposé de l'erreur; de même en philosophie expérimentale, ce ne sera pas l'expérience qu'on aura tentée, ce sera son contraire qui produira le phénomène qu'on attendait'.[60]

Moreover Diderot asserts that lines of questioning should not be abandoned lightly. The degree of vigour that one applies in an investigation should depend upon the degree of analogy between one's conjecture and the phenomenon that it is attempting to explain. If an idea as to how something might work is absolutely bizarre, it may merit only one test before being discarded, but if it seems more congruent with what we already know about nature, then the process of interrogation should be of greater longevity. This aspect of Diderot's approach is explained most clearly in *Pensée 42*, but elsewhere in *L'Interprétation de la nature* he alludes to the fact that one of the major problems with experimental philosophy is lack of appropriate wherewithal in following through with lines of investigation.[61]

While Diderot insists that experiments should not be abandoned lightly, he also emphasises that the philosopher must never remain too

59. *Pensées sur l'interprétation de la nature*, p.761.
60. *Pensées sur l'interprétation de la nature*, p.750.
61. *Pensées sur l'interprétation de la nature*, p.750.

attached to his own opinions. In other words, experiments must always be given their autonomy; if we are willing to accept positive results, we must also be prepared, after the experiment in question has been repeated a suitable number of times, to concede defeat.[62] Experiments serve the purpose not only of 'shoring up', but of 'cutting down' as well.[63] In the latter case the philosopher will need to return to the reflective stage to contemplate his ideas anew. This is an important point, for it is part of Diderot's insistence that the natural philosopher must never respond for nature, but rather adapt his own ideas to what is empirically discovered.

This of course, means invoking the third pillar of Diderot's method: reflection. The realm of reflection vis à vis Diderot may be considered in general terms as the realm of thought, the realm in which Man's intellectual faculties take the simple ideas furnished by experience and combine them to form complex ideas. The notion of reflection can be taken to incorporate the activity of rational philosophers, but it is also much larger than this; it refers in a general way to the process of thinking about what has been 'discovered' (or, we might say, thinking about the interrelation of the facts that we have 'uncovered') and making sense of it.

Reflections connect observations with experiments. They give order, meaning and purpose to the information collected by these empirical means. It is ultimately only through reflection that various phenomena can be connected to one another and, as Diderot reminds his reader in *Pensée 11*: 'L'indépendance absolue d'un seul fait est incompatible avec l'idée de tout; et sans l'idée de tout, plus de philosophie.'[64] Thus, reflection is the lynchpin of human knowledge, it is nothing without empirically collected information to contemplate, but without the process of contemplation this information would remain meaningless.

It is in this context that reflections must be considered to play a seminal role in determining the course of experiments to be performed. This is affirmed in *Pensée 29*, which claims that a suite of experiments is suggested by observations and the systematic ideas of rational philosophy.[65] This is not to say that one must be guided by a complete system of thought, but simply that the process of reflection upon the evidence at hand and the development of new ideas on the basis of this evidence is fundamental to deciding what experiments to undertake and how to do them. Experiments then provide confirmation (or contradiction) of the theory that reflection has formed on the basis of observations. The elements of Diderot's method thus become fully intertwined, as reflection emerges as an interpretation of observations and the results of experiments, as well as a means of suggesting the course of further research.

62. *Pensées sur l'interprétation de la nature*, p.750.
63. *Claude et Neron*, p.588.
64. *Pensées sur l'interprétation de la nature*, p.722.
65. *Pensées sur l'interprétation de la nature*, p.731.

Where *L'Interprétation de la nature* begins to emerge as truly novel and to stretch the parameters of Baconian empiricism is by virtue of the fact that under the idea of reflection Diderot subsumes the notion of hypothesising or 'conjecture'. It is worth remarking briefly here on the history of the notion of hypotheses in natural science. Perhaps the most famous statement with regard to the development and use of hypotheses in natural philosophy is Isaac Newton's oft quoted 'hypotheses non fingo', which is part of a statement coming towards the end of his *Principia mathematica*:

> But hitherto I have not been able to discover the cause of those properties [of gravity] from phenomena, and I frame no hypotheses; for whatever is not deduced from the phenomena is to be called an hypothesis; and hypothesis, whether metaphysical or physical, whether of occult qualities or mechanical, has no place in experimental philosophy.[66]

Without doubt, this statement cast a long shadow over the world of eighteenth-century philosophy. Many of Diderot's contemporaries were duly suspicious of the application of hypotheses in natural philosophy, and many natural philosophers rejected the notion outright. Pieter van Musschenbroek and the abbé Nollet, for example, were two vociferous opponents of hypothesising.[67]

Of course, we must carefully consider here what is meant by the term hypothesis. After all, for any twenty-first-century schoolchild, the word hypothesis plays a crucial role in the process of hands-on science learning. Must not the scientist have a hypothesis worked out before the experiment can be conducted? Is it not the hypothesis that helps us to know what it is we are looking for? To be sure, the answer to both of these questions is yes. But it was not hypothesis in quite this sense that Newton, van Musschenbroek or Nollet were talking about. In proscribing hypotheses from natural philosophy, all of the above-mentioned thinkers were essentially arguing that unsubstantiated speculation cannot be taken as a substitute for knowledge. Ideas deduced exclusively from abstract first principles, ideas that made no reference to empirical evidence, educated guesses, uneducated guesses and theories that were simply products of the unrestrained human imagination were not to be admitted as valid, and had no right to pass themselves off as telling us anything about the content of the 'real world'.

Diderot's position was not entirely dissimilar. True to his empirical form he too refused to place any stock in unwarranted speculations undertaken for lack of solid evidence. His point in defending the notion of hypothesising was merely that the natural philosopher should be able to think broadly and to exercise creativity as part of the process of

66. Isaac Newton, *Principia mathematica*, translated by Andrew Motte, edited and presented by Florian Cajori (Berkeley, CA, 1960), 'General Scholium', bk 3, p.547.

67. Mayer, *Diderot, homme de science*, p.116.

reflecting upon the facts before him, presuming that the fruit of such consideration was then tested on the anvil of experience. Thus hypotheses emerge as a potentially useful tool in the interpretation of nature, at least as long as they have a fixed and determinate idea, that they actually explain something, and that they do not invoke difficulties even trickier to explain than the original problem.[68] By way of giving examples of what he proposes by the term hypothesis, Diderot offers in *L'Interprétation de la nature* a series of conjectures. These include discussions about matters such as the nature of the celestial sphere, electricity, magnetism, solubility and elasticity. These conjectures fulfil a role similar to the 'Queries' section of Newton's *Opticks*. They do not necessarily bind the natural philosopher to anything, they are not taken to constitute actual knowledge, they simply provide the scientist with an occasion to meditate upon the potential implications of what research has uncovered. Thus, even if only by suggesting new directions or guiding the researcher down paths where something useful is accidentally discovered, Diderot felt that conjectures had a role to play. They involved for him no more than allowing the interpreter of nature to think laterally, to ruminate upon the unlikely or the possible, and thereby either clarify his position or perhaps attain insight which leads to a new discovery.

But Diderot's admission of hypotheses to the pantheon of tools at the disposal of the natural philosopher represents only part of the story. In a move that must have confounded most of his contemporaries, he asserted not only that conjecture was a key element of natural philosophising, but that the process did not take place on a purely rational level. He argues in *Pensées 30* and *31* that seminal to the interpretation of nature is what can be described as a 'feeling' for nature. This irrational element Diderot labels *pressentiment*. Described as a quality 'that is akin to inspiration',[69] *pressentiment* might best be understood by the modern reader as intuition (a term that Diderot avoided, as John Hope Mason speculates, on account of its religious connotations).[70] Inferring creativity, imagination and – for lack of a better term – gut instinct, intuition is utilised by the philosopher in combining reflections so as to form hypotheses. In other words, the scientist uses not only the rational faculties of induction and

68. See the *Encyclopédie* article 'Hypothèse', vol.8, p.418.

69. *Pensées sur l'interprétation de la nature*, p.732.

70. John Hope Mason, *The Irresistible Diderot* (London, 1982), p.58. It should be noted that Descartes spoke of intuition, but used the term to denote immediate knowledge of reality, an understanding of something that cannot be reduced and can only be regarded as axiomatic. Desmond Clarke describes it as a form of 'intellectual seeing' which seizes upon the 'unanalyzable datum of experience' (see Clarke, *Descartes' philosophy of science*, p.58-61). Thus intuition was a key element of what may be termed Descartes's philosophy of mathematics. There does exist the possibility then that Diderot refrained from using the term intuition not on account of its religious connotations, but in order to avoid confusion with a potential meaning vis à vis mathematics and its place within physical science.

deduction to guide the choice and structure of experiments, but also feeling. Diderot fleshes out this idea with a classical reference:

> La grande habitude de faire des expériences donne aux manouvriers d'opérations les plus grossiers un pressentiment qui a la caractère de l'inspiration. Il ne tiendrait qu'à eux de s'y tromper comme Socrate, et de l'appeler un démon familier. Socrate avait une si prodigieuse habitude de considérer les hommes et de peser les circonstances, que dans les occasions les plus délicates, il s'exécutait secrètement, en lui, une combinaison prompte et juste, suivie d'un pronostic dont l'événement ne s'écartait guère. Il jugeait des hommes comme les gens de goût jugent des ouvrages d'esprit, par sentiment. Il en est de même en physique expérimentale, de l'instinct de nos grands manouvriers. Ils on vu si souvent et de si près la nature dans ses opérations, qu'ils devinent avec assez de précision le cours qu'elle pourra suivre dans les cas où il leur prend envie de la provoquer par les essais les plus bizarres.[71]

Thus, not only does *L'Interprétation de la nature* place emphasis on sensory acquisition and reflection, but also on the capacity to move the mind in ways that are not always predictable or logically implied by what has gone before. But if logic is not necessarily inherent in the operation of *pressentiment,* this does not mean that the products of intuitive hypothesising are likewise irrational. Diderot is unequivocal about the fact that reliance upon the 'démon familier' must still end up developing notions that are clear and intelligible. As he states in *Pensée 40*: 'ce n'est pas assez de révéler; il faut encore que la révélation soit entière et claire. Il est une sorte d'obscurité que l'on pourrait définir, *l'affectation des grandes maîtres*. C'est un voile qu'ils plaisent à tirer entre le peuple et la nature [...] Hâtons-nous de rendre la philosophie populaire.'[72] Thus Diderot asserts that, however they are developed, the theories of the natural philosopher must be communicable.

This route is not without its pitfalls, of course. We may well ask how it is possible to begin with 'feeling' or sensing in a supra-linguistic fashion and arrive in the end at ideas that can nevertheless be expressed with the kind of clarity and distinctiveness emphasised by Descartes. Diderot does not provide an explicit answer to this. It may, however, be presumed that for him *pressentiment* is not a way of knowing but a way to knowing. That is to say that it could be viewed as a tool used as part of the process of discovery; it is not necessarily that things are only grasped intuitively, but that an intuition leads the experimental philosopher to try or see something that constitutes a discovery. This discovery (if it is going to be admissible to the body of Man's natural philosophical knowledge) must however be apprehended with a certain clarity and distinctiveness, and must of course also be amenable to being expressed with like clarity.

71. *Pensées sur l'interprétation de la nature*, p.732.
72. *Pensées sur l'interprétation de la nature*, p.747.

The fact remains however, that Diderot smuggles notions of intuition and instinct into the process of hypothesising. In so doing he strengthens his own metaphor of the experimental physicist being one who proceeds, at least in part, by smelling, touching and tasting. But Diderot's interpreter of nature also enters into a very immediate relationship with his subject. The act becomes a highly individual, personal and imaginative pursuit. The products of natural philosophy may need to be clearly expressed and widely communicable, but the actual interpretation of nature is a profession for which only a very few are equipped. This introduces Diderot's concept of the genius. On a number of occasions throughout *L'Interprétation de la nature* Diderot makes reference to the idea that true scientific progress is dependent upon the emergence and work of brilliant individuals. Essentially, the genius envisioned by Diderot must encapsulate all of the positive qualities that are outlined in *L'Interprétation de la nature*. He must be profound, have depth of insight, observe assiduously, perform experiments with exactitude, reason well and be able to draw the necessary linkages between phenomena and in so doing create a coherent picture of reality. In addition to all of this, Diderot's genius must, of course, be blessed with the gift of *pressentiment*. He must have an instinct for nature and its operations, be adaptable to its demands and allow the object of his investigation to work on him just as he works on it. For Diderot, Man is a part of nature, and cannot escape its bounds. Even though human attempts to establish order and meaning on the world inevitably separate Man from nature to some degree, Diderot employs the concept of the genius to try and ensure that this separation remains minimal. Only by remaining 'in tune' with nature, and keeping oneself open to its influence can the genius ever portray it in a remotely accurate fashion. There is an element of symbiosis here; the scientist brings himself to bear on nature, but equally opens himself up to its guidance.

Thus, while Diderot maintains that the pursuit of the experimental philosophy is open to all, he very definitely posits that the proper interpretation of nature is not. There is a strong sense of elitism which emerges with Diderot's conception of genius, a sense that recurs throughout his work. Much of *Réfutation d'Helvétius* (to be treated in greater detail in subsequent chapters) is devoted to defending the concept of genius against the assault of Helvetius's strictly egalitarian behaviourism. Diderot argues that education alone is not sufficient to produce genius, and that those who have most greatly contributed to the advancement of human knowledge were separated from the common run of men by their special and innate abilities. Newton, according to Diderot, was born, not made. In maintaining such a stance, the *philosophe* had gone from stretching the bounds of Baconian empiricism to breaking them entirely, for it was Bacon who said: 'For our method of discovery in the sciences more or less equalises intellects, and leaves little opportunity for superiority, since it achieves everything by most certain rules and forms of proof.'[73]

It is not to be wondered at that Diderot defends the concept of genius so vigorously. It is a notion crucial to both his scientific and his artistic thinking.[74] Without the notion of the genius, Diderot's philosophy of science would be severely lacking. His natural philosophy requires the employment of imagination, creativity, instinct and a process of passing between highly personal introspection and concrete activity. Of course, one must be wary of overemphasising these elements of Diderot's thought. It would be easy to get carried away and all but forget the more rational, traditional elements of his natural philosophy. As important as it was to Diderot that the interpreter of nature could exercise a highly personal and perhaps ultimately indescribable finesse, he knew that logic, mathematisation, clear expression of ideas and a strong connection to what is concrete were just as important to the progress of human knowledge. Little wonder then that Diderot perceived the genius as being an exceptionally rare breed, and that this loomed for him as one of the greatest obstacles to human progress.[75]

What emerges from *L'Interprétation de la nature* then is a bold and incisive picture regarding what it means to be an interpreter of nature. It casts the discipline of experimental physics into a unique and often thought-provoking light. It is unsurprising therefore that modern commentators have interpreted it in a wide variety of ways. In addition to noting the work's aforementioned and readily apparent Baconianism, many scholars have placed stress upon Diderot's debt to the work and thought of Newton.[76] The emphasis here usually tends to be on the latter's *Opticks*, which may be regarded, alongside the works of Bacon, as one of the seminal texts of early experimental science.[77] The primary roles accorded in this work to experience, the inductive method, carefully controlled and interpreted experiments and even the process of conjecture are all represented in *L'Interprétation de la nature*. In contrast however to this pro-Newtonian reading of the situation, there is yet another set of historians that have sought to play up the role of introspection and

73. Francis Bacon, *The New Organon*, ed. Lisa Jardine and Michael Silverthorne, bk 1, Aphorism 122, p.95.

74. Of course, many scholars will be more familiar with Diderot's notion of the genius in terms of his aesthetic thought. For Diderot, the idea of genius was inextricably connected to the process of both artistic expression and that of natural philosophical discovery. A good, and classic introduction to this aspect of the *philosophe*'s thought is Jacques Chouillet, *La formation des idées esthétiques de Diderot*, 1745-1763 (Paris, 1973).

75. *Pensées sur l'interprétation de la nature*, p.719.

76. Regarding the influence of Newton upon *De la nature*, see Anderson, *Diderot's dream*. Anderson claims that while *L'Interprétation de la nature* is undoubtedly a mixture of different strains of thought, 'its overall structure is Newtonian' (p.26).

77. For more on the place of Newton's *Opticks* in the rise of modern science, see A. Rupert Hall, *All was light: an introduction to Newton's Opticks* (Oxford, 1993).

reflection in the work, and on these grounds have argued that one of the primary influences is the thought of Descartes.[78]

Each of these interpretations is not without merit, but *L'Interprétation de la nature* is at the very least a work of synthesis, an attempt to harmonise the legacies of Descartes and Newton, of rationalism and empiricism, of the necessity for abstraction and the desire for loyalty to that which is concrete and particular. Most scholars, whatever their other commitments, have acknowledged this to one degree or another. Some have gone even further however and argued that with *L'Interprétation de la nature* Diderot was not simply harmonising but actually trying to transcend the Newtonian–Cartesian conflagration. Wilda Anderson sees Diderot's reliance on Bacon as a way of accomplishing this.[79] Laurence Forno plays up Diderot's respect for the underlying unity of nature, his advocacy of the need for symbiosis between natural philosopher and subject, and his reliance on the notion of *pressentiment* to suggest that *L'Interprétation de la nature* is characterised by a 'cosmic mysticism' which allows Diderot to transcend the prevailing dichotomies of the day.[80] While not necessarily subscribing to the precise terms of Forno's interpretation, his basic stance of perceiving *L'Interprétation de la nature* to be a work of transcendence makes sense. The greatest danger that arises when discussing Diderot's philosophy circa 1753 stems from trying to classify it simply according to various categories of seventeenth-century thought. Its inspiration, tone and content are rather a true mixture of Bacon, Locke, Descartes, Newton and, not to be underestimated, Diderot's own ideas.[81] If nothing

78. Aram Vartanian, *Diderot and Descartes: a study of scientific naturalism in the Enlightenment* (Princeton, NJ, 1953). Vartanian argues that scholars have grossly exaggerated Diderot's loyalty to Bacon (p.136), and goes on to claim that 'in actual elaboration of scientific doctrine Diderot was only too often given to bold speculation, sweeping hypotheses and deductive reasoning' (p.137). On this basis Vartanian characterised *L'Interprétation de la nature* as firmly bearing the stamp of Cartesianism. While there is an undeniable Cartesian influence apparent in *De la nature*, Vartanian's rather off-handed denigration of Diderot (as 'only too often given to bold speculation' etc.) seems flawed, or at least improperly grounded. Diderot did on occasion indulge in speculation, hypothesisation and indeed deductive reasoning, but each of these things needs to be properly understood within the framework of his scientific methodology.

79. One of the problems with this reading is that it devotes very little space to the matter of mathematics, and tends to accept quite willingly the thesis (criticised in chapter 1) that Diderot opposed the idea that the discipline was a useful tool in the investigation of nature (see Anderson, *Diderot's dream*, p.17). As a result, the tension that Anderson sees as being transcended by Diderot is not so much empiricism versus rationalism (which here could be taken to subsume the mathematisation of nature promoted by Descartes), but rather empiricism versus the intuitive, creative aspects of natural philosophising. In fact, *L'Interprétation de la nature* must be read as a blending of (and an attempt to transcend the pitfalls nascent in) all three approaches: the empirical, the rational and the intuitive.

80. Laurence Forno, 'The cosmic mysticism of Diderot', *SVEC* 143, p.113-40.

81. Chapter 1 has also demonstrated the impact of Buffon on *L'Interprétation de la nature*, and shortly we shall see that the work engages with other contemporary thinkers as well, such as Maupertuis.

else, *L'Interprétation de la nature* is a perfect example of the fluidity of the mid-eighteenth-century French intellectual landscape in which individual natural philosophers, such as Diderot, were to a large degree left to their own devices in terms of evaluating and applying the varieties of intellectual currency that existed, and contriving what seemed to them the most reasonable approach to studying nature.

A perfect illustration of this fluidity can be obtained by contrasting the viewpoint of Diderot with that of D'Alembert. Although both men valorised Newton and perceived themselves to be standard-bearers for the advancement of Newtonian natural philosophy within France, the actual content of their respective scientific methodologies is remarkably different. Regardless of his rhetoric, D'Alembert's practice of natural philosophy tended to downplay the value of experiment and observation.[82] This is not to say that he ruled out the discipline of experimental physics altogether, but he did tend to pass over it rather lightly, and to regard it with no small measure of disdain. His *Essai sur les Éléments de philosophie*, for instance, provides an overview of what he considers to be the most significant realms of natural philosophy, and tellingly manages to overlook experimental physics as a distinct category. This, on its own, does not necessarily mean much, and indeed D'Alembert does devote some attention to the subject of 'physique expérimentale' in his discussion of 'physique générale'. What does strike the reader however, is the manner in which D'Alembert refers to knowledge that has been acquired primarily by means of observation and experience, calling it 'la physique vulgaire et palpable' and claiming that we can 'réserve pour l'expérience la nom de physique occulte'.[83] The only real use of experimental physics according to D'Alembert's conception is to examine the general properties of bodies that experience alone can measure, and of which it alone can determine the effects.[84] An example of this might be Galileo's experiments with ball and plane. Other examples could include trials that, in an eighteenth-century context, were much more along the lines of chemistry, and which were aimed at doing things like discerning the density of a given substance, the hardness of a particular object, or the effect of subjecting some type of body to heat.

That which truly marks the thought of D'Alembert is a distinct preference for the analysis of the world in a highly abstract and mathematical manner.[85] Favouring a purely quantitative approach to under-

82. Thomas Hankins has aptly observed that D'Alembert, despite persistent eulogising of Isaac Newton, virtually ignored the *Opticks* and was also quite oblivious to parts of the *Principia mathematica* that emphasised experimental evidence and observation of common phenomena. See Hankins, *Jean D'Alembert*, p.24.

83. Jean D'Alembert, *Essai sur les Eléments de philosophie*, in *Œuvres de D'Alembert*, vol.1 (Paris, 1821), p.337.

84. D'Alembert, *Eléments de philosophie*, p.343.

85. Hankins, *Jean D'Alembert*, p.3.

standing nature, he presumed that the 'truths' of physics could be derived from mathematical generalisations. Hypotheses were only relevant insofar as they could be quantitatively checked against phenomena. Thus, D'Alembert argued that: 'L'étude de la nature est celle des propriétés des corps; et leurs propriétés dependent de deux choses, de leur movement et de leur figure. Ainsi les sciences qui s'occupent de ces deux points, c'est-à-dire la méchanique et de la géométrie, sont les deux clefs indispensablement necessaries de la physique.'[86] In other words, for D'Alembert, the general properties of bodies and the effects of the action which they exercise upon one another were the first objects offered to the scientist in the study of nature. These properties and effects were seen to be virtually self evident, resulting from the nature of bodies themselves.[87] Presumably, it was not a complicated or contentious process in the eyes of D'Alembert to immediately abstract from experience such essential items, and he suggests as much when he argues that simple observations are easily reduced to basic laws.[88] In this way D'Alembert hoped to have legitimated the method of the mechanician/philosopher, who could begin with the fundamental quantifiable properties of bodies and from these proceed with geometry and algebra to figure out the true laws of nature. Thus, although D'Alembert, like Diderot, claimed to have little patience for metaphysics or systematic, rational philosophising, he had still less appreciation for descriptive science. His philosophy retained a very significant element of deduction in its approach to establishing knowledge of the world. To borrow the terminology of modern philosophy of science, it might even be described as representative of the 'hypothetico-deductive model' of scientific reasoning. This model of science advances a methodology by which observation gives rise to axioms (which are usually mathematical), and the main business of the real scientist is to deduce from these axioms what must be, check this against reality and enshrine the results, if positive, as law.

As has been made apparent from chapter 1, deduction, as in the guise of mathematics, was not anathema to Diderot's conception of natural philosophy; it was, at the very least, an example of the best means by which to proceed with logic, clarity and rigour. Nevertheless, it could not in itself lead to new knowledge of the world, and consequently could not provide the basis for drawing conclusions about nature. As has already

86. D'Alembert, *Mélanges de literature: d'histoire, et de philosophe*, 4 (Paris, Briasson, 1753), p.153.

87. D'Alembert, *Eléments de philosophie*, p.342.

88. This was, of course, a contentious process in Diderot's eyes. He would claim here that D'Alembert is commencing his analysis of reality at a point that is already unjustifiably high up the ladder of induction – in other words, that D'Alembert is naively taking too much for granted. As much as anything else, this kind of methodological discrepancy highlights just how far removed from one another the philosophical conceptions of Diderot and D'Alembert could be.

been suggested above, the fundamental movement of natural philosophy for Diderot ultimately rested on induction. A reading of *L'interprétation de la nature* makes it absolutely clear that moving from observed reality to theory was a cornerstone of Diderot's natural philosophical methodology. This notion is firmly entrenched throughout the canon of Diderot's work. In the article 'Encyclopédie' for example, Diderot states that 'toute notion est ou de sensation ou d'induction'.[89] Elsewhere he stresses that the work of science must proceed from the 'petite' to the 'grande', rather than from the 'grande' to the 'petite', as in *Le Rêve de D'Alembert* where it is stated that good scientific method involves finding out about larger bodies by studying small ones.[90] In other words, Diderot insists on working up from particular instances to larger generalisations, rather than starting 'at the top' of the chain of reasoning and working down to actual phenomena. Crucially he also stresses the extraordinary, potentially infinite complexity of this entire process.

This sounds eminently reasonable of course, but the fact remains that a problem which has long preoccupied philosophers of science is that induction is not by any means a straightforward process. The main difficulty is determining exactly when one has collected enough 'facts' upon which to base a conclusion that can be regarded as valid. Many, such as the Scottish philosopher David Hume, have asserted that we can never be in possession of enough evidence to warrant the establishment of a truly valid conclusion. Roselyne Rey has argued that Diderot's position is not dissimilar, and that rather than establishing a basis for natural philosophy *L'Interprétation de la nature* seriously undermines any hope for proceeding by inductive reasoning. According to this interpretation, Diderot questions the process of moving from the particular to the general as well as the degree to which a particular object can be properly identified in the first place.[91] His method, according to Rey, is not to be stuck on any particular method, particularly when it comes to dealing with induction.[92] This is, in many respects, a compelling and understandable appraisal. Diderot does not, either in *L'Interprétation de la nature* or elsewhere, develop any hard and fast rules of induction.[93] As has already been observed on a number of occasions in this book, Diderot's work is characterised by a high degree of indeterminacy and

89. 'Encyclopédie', p.383.

90. *Le Rêve de D'Alembert*, p.89.

91. Roselyne Rey, 'Dynamique des formes et interprétation de la nature', in *Recherches sur Diderot et l'Encyclopédie*, vol.11 (1991), p.58.

92. Rey, 'Dynamique des formes', p.53.

93. Most philosophers of science would agree in any case that the development of such rules is a prohibitively difficult, if not impossible task. Rey underlines this by stating that Diderot considered it insufficient and actually anathema to the progress of science to try and institute rules for passage from the particular to the general. See Rey, 'Dynamique des formes', p.53.

suspicion regarding the acquisition of 'true' knowledge about the world. Nevertheless, he did not seek to abandon the process of induction entirely, for admitting that would have been tantamount to forsaking the scientific enterprise and indeed all claims to any worthwhile knowledge.

How then might Diderot be perceived to come to terms with the problem of induction that rests implicitly within his scientific methodology? To a certain extent Diderot's elucidation regarding the role of reflection can be invoked here. In developing theories or hypotheses on the basis of what we observe in nature, we are moving from the level of the particular to the more general. Although Diderot acknowledges that this movement is not infallible, he attempts to bolster its acceptability by insisting that the philosopher is continually shuttling back and forth between observation and reflection, that he employs the ambiguous attributes of creativity and intuition, and that he checks his hypothesis against experience not simply by means of renewed observation but also by contrived experiments. Perhaps most importantly, however, the problems inherent in induction ultimately point towards a crucial aspect of Diderot's natural philosophy: his appreciation for the fact that science is an ongoing enterprise. This is absolutely seminal to his vision of natural philosophy, and is a point to which we shall return shortly.

As a further part of the response to the means by which Diderot proposed moving from the particular to the general in a way that might be considered useful, we must look beyond the confines of *L'Interprétation de la nature*. That is to say, there emerges from his canon of work a set of evaluative standards that assist and elaborate upon that which is contained in his classic treatise on methodology. Some of these items may be inferred in *L'Interprétation de la nature*, but others are not. Furthermore certain of these standards are explicitly stated by Diderot at one time or another, while others are implicit in the manner that he either broaches certain substantive issues within the realm or physics or takes a stance regarding the nature of a particular scientist's work. *De la nature* is, after all, largely a work of prescription, rather than a tract on physics. In other words, in considering the nature of Diderot's natural philosophical methodology, we must look at not only what he preaches, but at what he practises. The result might be conceived not simply as a way of getting around the problem of induction, but more generally as the set of methodological rules that emerge from Diderot's natural philosophical work as a whole. These criteria can then be taken as the evaluative standards and warrants for the acceptance of a theory that underwrite his natural philosophy.

One of the most prominent lessons to emerge from Diderot's canon of natural philosophical work as a whole is his desire for rigour. This may seem commonplace, or even common sense, but it is worth emphasising if only to re-affirm that the intuitive, creative aspects of the natural philosophical process as outlined in *L'Interprétation de la nature* do not overrule reason. A number of Diderot's works stress the importance of ensuring that we are disciplined about drawing conclusions from the evidence at hand. This is significant, for Diderot clearly felt that much of the philosophy of his day was conducted on the basis of less than solid reasoning. As he states in *Claude et Neron*: 'rien de plus difficile que de ne tirer de l'expérience ou de l'observation que des conséquences rigoureuses'.[94] It will be remembered that for Diderot, mathematics, and geometry in particular, emerged as the paradigm of right reasoning. More than classical logic, it encapsulated and demonstrated the best means of linking ideas, drawing conclusions and establishing the necessary consequences of any given proposition. The best example of this trend is *Plan d'une université*, which posits that geometry is the best cure for the ignorance, superstition and wrong-headedness of a people.[95] But Diderot's 'geometric' mode of reasoning is apparent in much of his earlier work as well. *L'Interprétation de la nature* compares the process of the experimental philosopher linking phenomena to the work of the geometer, in that both seek to establish all of the steps that lie between two ideas.[96] Even if it were not always possible in the experimental sciences to elucidate all of the intermediary phenomena that lie between two others as it is in geometry,[97] Diderot felt it incumbent upon the natural philosopher to aim for such clarity. The implication of this is that preference should be given to scientific theories in which concepts and phenomena are most rigorously related, as opposed to those which may rest upon empirical data but are unable to demonstrate these connections with such necessity.

More interesting perhaps is the fact that there is in Diderot's work something that may be labelled a fondness for theoretical elegance. In this he strays a little from the straight and narrow path of inductive reasoning, for there is a sense that emerges on occasion from Diderot's work that, all other things being equal, the superior theory is the one that is, for lack of a better term, more aesthetically pleasing. This is a methodological preference that does not follow strictly from experience. Indeed, there does not seem to be any formal, logical justification for it whatsoever. Nevertheless Diderot demonstrated a very real preference for the elegant over the messy, the simple over the complex. There emerges the presumption in much of his work that neater and simpler theories have a superior explanatory capability.

This manner of thinking appears as early as *Lettre sur les aveugles* where, as we have seen in chapter 1, Diderot observed that scientific results were more or less certain depending on the relative complexity of the hypoth-

94. *Claude et Neron*, p.589.
95. *Plan d'une université*, p.773.
96. *Pensées sur l'interprétation de la nature*, p.720.
97. *Pensées sur l'interprétation de la nature*, p.752.

esis; the simpler the hypothesis, the stronger the force of the argument, the greater the number of suppositions incorporated within the hypothesis, the more tenuous the conclusion becomes. The presumption is that explanations consisting of more elements contain a greater number of suppositions, and therefore a greater number of junctures at which the theory can go amiss. Simpler theories, by contrast, are tighter, and leave less room for error.

This tendency is manifested much later in *Sur la cohésion*. As has been mentioned, the primary thrust of this work was to confound the intentions of many post-Newtonians (such as Clairaut) who sought to resolve problematic elements of attraction theory by adding various layers to the relatively elegant inverse square law.[98] It should be noted that Diderot was not so dogmatic as to rule out automatically the possibility that the inverse square law of attraction could be susceptible to some variation; he admitted that there are some cases (such as with regard to the refraction of light) where nature does seem to vary or modify its behaviour according to circumstances. Furthermore, as we have seen, his primary objection to those who would complicate Newton's law had less to do with the inherent nature of their theories than with the fact that these theories were not in accordance with phenomena. Still, in reading *Sur la cohésion* one acquires a strong sense of Diderot's basic distaste for the idea of altering the law by making it more complex. The very tone of the work is tainted with the prejudice of one who laments the entire notion of unnecessarily complicating a matter that should be simple. He speaks of the idea of developing a parallel law in order to account for phenomena as being 'une complication de termes embarrasante et même un peu bizarre'.[99]

This preference of Diderot for the simple over the complex was quite commonplace. It is essentially a predilection to invoke Ockham's razor (that is best which is simplest and works), although it might also be seen to bear a relation to Maupertuis' principle of least action, which states that nature always favours the easiest, most direct route from point A to point B. This is not to say that Diderot was necessarily influenced by Maupertuis in this respect, and in any case there is no firm evidence to confirm that this was the case. It should also be noted that for Maupertuis the principle of least action contained a metaphysical element in that it was used to establish the existence of God on the basis of the elegant and efficient construction of nature and nature's laws, whereas for Diderot the preference for simplicity was a tool in the elaboration of theory, or perhaps better, a philosophical prejudice.[100] One might also connect Diderot's thoughts on the matter to those of Newton, whose first Rule of

98. See chapter 1, p.19 and n.33.

99. *Sur la cohésion*, p.10.

100. David Beeson, *Maupertuis: an intellectual biography*, *SVEC* 299 (1992), p.271.

Reasoning in Philosophy stated 'We are to admit no more causes of natural things than such as are both true and sufficient to explain their appearances'.[101] This statement is well aligned with Diderot's thoughts on unravelling the complicated web of causes and effects in which he perceived so many of his contemporaries to be getting caught. Fundamentally, he advocated that for the natural philosopher the best means of proceeding with an explanation of phenomena was to seek to reduce the number of causes responsible for a given set of effects. Another way of putting it is that the philosopher should try to multiply the number of effects ascribed to a given cause.[102] Diderot's basic presumption then is that nature may be complex, but in our understanding it should be made relatively simple. This is, according to Diderot, the grand movement of science. Such a sentiment is reflected in *L'Interprétation de la nature*, which delivers the injunction that philosophers must seek to reduce 'un phénomène très compliqué à un phénomène simple'.[103]

Between *De la nature* and the rest of his work then, Diderot offers the natural philosopher a broad and well-defined platform for interpreting nature. The careful linkage of observation, reflection and experiment, the development of hypotheses, the guiding hand of instinct or feeling for nature, the rigourousness of geometric reasoning applied to the induction of theories from painstakingly observed phenomena and a consistent drive to render the complex into the simple are all part of a network of intellectual tools Diderot places at the disposal of the scientist. Yet despite the plethora of equipment that Diderot places at the disposal of the natural philosopher, and despite the undoubted boldness of his support for employing imagination and conjecture as part of the process of interpreting nature, there is a fundamental wariness that emerges from Diderot's work. This wariness stems at least in part from Diderot's conviction that there are and will always remain details of nature's operation that are veiled from human attempts to lay reality bare. He frequently asserted that nature often operates according to 'imperceptible degrees'. In *Pensée 37* of *L'Interprétation de la nature* he states: 'La nature est opiniâtre et lente dans ses opérations. S'agit-il d'éloigner, de rapprocher, d'unir, de divisier, d'amollir, de condenser, de durcir, de liquéfier, de dissoudre, d'assimiler, elle s'avance à son but par les degrés les plus insensibles.'[104]

Elsewhere, in *Nouveaux mémoires sur différents sujets de mathématiques*, Diderot argues against D'Alembert's position regarding the calculation of odds in games of chance partly on account of the fact that when the game is considered to be taking place in the real world we must count on

101. Newton, *Principia mathematica*, p.398.
102. *Claude et Neron*, p.588.
103. *Pensées sur l'interprétation de la nature*, p.753.
104. *Pensées sur l'interprétation de la nature*, p.744.

the effects of secret, inappreciable physical causes.[105] In the article 'Encyclopédie' he claims that 'L'univers ne nous offre que des êtres particuliers, infinis en nombre, et sans presque aucune division fixe et determine [...] tout s'y enchaîne et s'y succède par des nuances insensibles'.[106] This presumption of operations taking place in nature which are beyond Man's abilities of detection presents an obvious and major obstacle to the natural philosopher. There is the sense in reading Diderot that regardless of how fine the instruments or how expansive Man's knowledge, nature, with its perpetual vicissitudes and degrees of operation so minute as to be insensible, will always remain at least one step ahead of its pursuers.

Thus there is an understandable conservatism that pervades almost all of Diderot's natural philosophical writing. He often indulges in potentially fruitful speculation, but invariably his works conclude on a hesitant, almost uncertain note. This might have been a reflection of his unease with the material at hand. Then again, many with far less knowledge than Diderot were quite content to establish firm and unwavering positions. Moreover it is not as though Diderot himself was without positions. His wariness of proclaiming a scientific truth with absolute confidence seems to stem rather from his view on the fundamentally tenuous nature of the relationship between Man and nature.

Though in terms of experimentation he felt it perfectly acceptable to try anything at least once, one of the strongest lessons to emerge from Diderot's thoughts on the experimental philosophy is that in terms of assembling theories it is much better to admit ignorance than to build up a system on the basis of unfirm or unconfirmed knowledge. He is, in a sense, very defensive, very aware of the inadequacies which pervade any proclamation of truth. Even in cases where Diderot feels relatively sure of his foundations, he refuses to commit himself entirely. Most of his works end with the injunction to perform more experiments, to pursue further investigation. Rather than coming to a firm and specific position he tends to offer some general conclusions and then indicate the prospective path for further research. While some may have considered this a weakness, Diderot undoubtedly viewed it as a tremendous strength. For him there was a certain valour to be found in the act of 'supposing' rather than 'affirming'.[107]

Caution is therefore the watchword that can be used to characterise much of Diderot's science. This is ironic, since he is usually acknowledged primarily for pushing the boundaries of eighteenth-century thought, and for indulging in imaginative if remarkably insightful flights of speculation such as *Le Rêve de D'Alembert*. Here, perhaps more than anywhere else,

105. *Sur les probabilités*, *OC*, vol.5, p.32.
106. 'Encyclopédie', p.408.
107. *Les Bijoux indiscrets*, *OC*, vol.1, p.589.

Diderot the philosopher intrudes upon and enlightens Diderot the scientist. His basic caution stems ultimately from a profound and subtle understanding of how epistemological matters underwrite all aspects of natural philosophy.

In this sense Diderot was very much in the mould of certain Dutch scientists of a slightly earlier generation who downgraded the status of what many contemporaries took to be 'certain knowledge', inferring that such a concept is not always helpful in natural philosophy. Perhaps most notable in this respect was Christiaan Huygens (1629-1695), one of the Continent's most incisive critics of Newtonianism, who advanced the notion that probability is almost as good as proof.[108] Likewise, Willem Jacob 's Gravesande (1688-1742) proclaimed in a fashion similar to Hume that the axiom of the uniformity of nature (so essential to the scientific enterprise) is not logical, but pragmatic.[109] Meanwhile Musschenbroek, inventor of the Leyden jar (a device for storing and transporting electricity) and a nominal Newtonian,[110] cast as 'necessary assumptions' much of what was taken by other Newtonians as knowledge.[111] While Diderot, like Gravesande and Musschenbroek, did much of his work comfortably within the structure created by Newton, he had a more critical and balanced view of this structure than many of his contemporaries. Diderot was probably more circumspect about the entire notion of truth than most scientists of his day. There emerges within his canon a veritable decline in the concept of absolute certainty, and a rise in that of probability.[112]

Thus, Diderot's methodology might be characterised in terms of a softening or tempering of Newtonianism. He eagerly accepted much of the Newtonian method and certainly preferred it to that of Descartes, but he also attempted to go beyond the parameters sketched out by the Englishman in the *Principia*. Diderot undoubtedly regarded the *Principia* as a work of genius, and accorded some value to the 'Newtonian style' of

108. Cassirer, *Philosophy of the Enlightenment*, p.60. For more on Huygens, see Joella G. Yoder, *Unrolling time: Christiaan Huygens and the mathematization of nature* (Cambridge, 1988). Very appropriately, for one who insisted upon the discussion of scientific truth in terms of relative probability rather than absolute certainty, Huygens was one of the first mathematicians to publish a work on the calculation of probabilities (Yoder, p.2).

109. Dieckmann, in *Isis*, p.251-52.

110. The Leyden Jar was actually invented by a Polish cathedral dean, Ewald von Kleist, but as he kept secret the process for construction of the new machine, the accolade of inventor eventually went to Musschenbroek. Regarding Musschenbroek's approach to natural philosophy, Voltaire proclaimed that he was 'naivety itself and loved Truth with a child's openness'. See Patricia Fara, *An Entertainment for angels: electricity in the Enlightenment* (Cambridge, 2002), p.51-52.

111. Dieckmann, in *Isis*, p.251-52.

112. This, of course, bears a relation to his interest in the mathematical discipline of probabilities, over which, as we have seen in chapter 1, he maintained strong differences of opinion with D'Alembert, who was in such matters more bound to the concept of absolute certainty.

natural philosophising contained therein, but his own vision of how knowledge of the world is acquired was much more closely related to the method exhibited by the *Opticks*. That is, Diderot sought on the one hand to liberate the natural philosopher from the strictures of quantification and also downgraded or at least put into a more humble perspective the veracity of key propositions. On the other hand, the means of investigation advocated by Diderot were greater in extent than those of Newton (that is, through the explicit inclusion within Diderot's methodology of factors such as *pressentiment*, creativity and flexibility) even if the conclusions reached by these means were not as strong, indelible or unequivocally expressed.[113]

These methodological predilections can now be related to Diderot's thoughts on the system of the world – that is, the physical structure of the universe and the nature of its composition. Here, just as with regard to his methodology, Diderot's thought betrays a shift from the Cartesianism of his early education towards and ultimately beyond a vaguely Newtonian paradigm. Although, as we have seen, it is not known exactly when Diderot first became exposed to the thought of Newton, it is clear that by the 1740s he had become well acquainted with many aspects of the Newtonian view of the universe and its operation. More importantly, he was generally accepting of this view. Much of Diderot's early work on physics, though it may have been more inclined towards physico-mathematics than towards experimental physics, dealt directly with issues arising from the *Principia mathematica* and the *Opticks*. As we have seen in chapter 1, *Mémoires sur différents sujets de mathématiques* makes frequent reference to Newton and even attempts to refine certain elements of Newton's work. Even by this early date, the fundamentals of the Cartesian philosophy were obsolete for Diderot; there was for him no consideration of the plenum, impulsion, vortices or any other significant aspect of Descartes's cosmology.

It was the nature and behaviour of matter at the most basic level however that was the central concern of Diderot's physics. Here Diderot's vision of the functioning universe made its most significant departure from the Cartesian world-view. This rejection of Descartes's view of matter took place on several fronts. First, Diderot rejected the postulate that matter was indefinitely divisible, arguing that this was a philosophical impossibility. Diderot believed, as is demonstrated in *Pensée 36* of *L'Interprétation de la nature*, that there must exist some manner of ultimate particle.[114]

113. Herbert Dieckmann has posited, for instance, that Diderot's notion of experiment diverged from that of Newton primarily in that its purpose was seen to be more than the simple reduction of phenomena to quantitative relationships (although, to reiterate the argument of chapter 1, such reduction was not exactly proscribed by Diderot either, and it played a very active role in his natural philosophical outlook). See Dieckmann, in *Isis*, p.252.

114. *Pensées sur l'interprétation de la nature*, p.738-44. This is the fifth and longest of

It must be noted, however, that this point was not particularly crucial to Diderot's understanding of matter. More significantly, he never made any dogmatic attempts to defend it. This was appropriate, since Diderot's rejection of the infinite divisibility of matter may be taken as an example of pure rationalism at work in his physics. There could not possibly have been any empirical demonstration as to the truth of the matter either for Diderot or any of his contemporaries; his assertion that matter could only be broken down to a certain point was the result of purely deductive reasoning. Well aware of the fact that both sides were on highly speculative ground, Diderot was appropriately wary of being drawn too far into this controversy. Though he objected to Cartesian ideas on the issue, his objection was rather timid and expressed in terms that are not inordinately bold. Significantly he also refused to speculate about the details of the ultimate structure or shape of the most fundamental particles of matter. Relative to other aspects of his matter theory, the divisibility issue seems not to have been very important to Diderot.

A far more significant and general characteristic of Diderot's physics was his rejection of the Cartesian matter and motion hypothesis. This notion asserted that the behaviour of all physical phenomena could be explained purely with reference to inert, homogeneous matter moving as a result of direct contact with other bodies. As Desmond Clarke explained:

> There are metaphysical and methodological constraints on what Descartes is willing to accept as a possible explanation, for he is uncompromisingly reductionist in his intolerance of theoretical entities. The demand that explanations ultimately reduce to descriptions of contact action between small particles of matter is aligned with a significant Cartesian dependence on mechanical models as an essential component in any scientific explanation.[115]

As will have been apparent from the survey of Diderotian methodology performed above, such a conception was too restrictive and too abstract for Diderot, who argued repeatedly that there was much more to the physical stuff of the universe than the relatively narrow and simplistic Cartesian theory would allow.

To a large degree however, Diderot's objection to the matter and motion hypothesis was due to its inability to properly account for phenomena. Following Buffon and Maupertuis, he perceived that exclusively mechanical models of the universe were insufficient for explaining the entire operation of nature. For Diderot, the matter and motion hypothesis was confounded by experience; other elements, and by extension a more subtle, complex idea of what constitutes matter were required in order to properly interpret the physical world.

Diderot's conjectures. In it he speculates on the nature of elastic bodies and enters into discussion of what modern science might term the molecular structure of such bodies.

115. Clarke, *Descartes' philosophy of science*, p.13.

One of the most obvious aspects of Diderot's rebuttal of the matter and motion hypothesis was his assertion that matter is heterogeneous. Quite simply, he did not perceive how the manifold productions of nature could result if matter was purely homogeneous. If the differences in individual pieces of matter were only of quantity, and not of quality, then the world would have none of the richness that it clearly does. Although he did not put the argument quite this way, it is as though Diderot is arguing that to posit homogenous matter as the sole source of the universe's diverse productions would be like arguing that something can come from nothing, and that making such an assertion would be to completely violate logic. For Diderot, the world that we experience is precisely the result of essential differences between the particles that comprise reality. As *Pensée 58* of *L'Interprétation de la nature* argues:

> Il me paraît aussi impossible que tous les êtres de la nature aient été produits avec une matière parfaitement homogène, qu'il le serait de les représenter avec une seule et même couleur. Je crois méme entrevoir que la diversité des phénomènes ne peut être le résultat d'une hétérogénéité quelconque. J'appellerai donc *éléments* les différentes matières hétérogènes, nécessaires pour la production générale des phénomènes de la nature; et j'appellerai *la nature* le résultat général actuel, ou les résultats généraux successifs de la combinaison des éléments. (*italics* are Diderot's)[116]

This was a point that Diderot made many times throughout his career, but perhaps most fervently in *Principes philosophiques sur la matière et le mouvement* (1770). We shall investigate this tract in greater detail in the next chapter, but for now, it suffices to say that the primary thrust of *Sur la matière et le mouvement* is to argue against the vain presumptions of rationalistic philosophers who posit that matter is homogeneous.

But Diderot's desire to investigate the universe on a level deeper than the simple mechanistic philosophy of Descartes would allow stemmed from more than just his rejection of the notion that matter might be homogeneous. Another strand of his attack pertained to ideas about the nature of motion, or more specifically ideas about the motions and activities ascribed to or associated with matter. The most significant departure of Diderot from the conception offered by Decartes was his acceptance of Newton's theory of attraction. Of course, the propositions that matter is homogeneous and that it is capable of being moved by a force acting at a distance are not at all mutually exclusive. It may be said that the view offered by Newton himself was essentially mechanistic and mathematical, and that it took little or no account of the heterogeneity principle that Diderot so frequently emphasised.[117] What Newton's the-

116. *Pensées sur l'interprétation de la nature*, p.767.

117. The essentially mechanistic character of Newtonian philosophy is firmly expressed by Jean Ehrard, who states that'On comprend que Newton ait été célébré plus tard par Laplace comme le tenant du plus absolu mécanisme.' Jean Ehrard, *L'Idée de nature en France dans la première moitié du XVIIIe siècle*, vol.1 (Paris, 1963), p.127-28.

ory of gravity did do for Diderot was draw him even further away from the physics of Descartes. In accepting that matter might be moved on account of forces acting at a distance, no room was left in Diderot's world-view for the Cartesian picture. Even Diderot's mechanics, the most basic level at which reality might be considered, were expressly Newtonian rather than Cartesian.

While accepting of Newton's inverse square theory of attraction, Diderot remained an agnostic regarding the cause of attraction itself. The best demonstration of this is in *Sur la cohésion*, where Diderot states that: 'Ces deux forces ne sont pas l'attraction même, mais des effets de l'attraction: car j'appelle attraction l'effort que fait le corps attirant pour faire mouvoir le corps attiré, et je regarde comme l'effet de l'attraction la force avec laquelle le corps attiré est mû en vertu de cet effort.'[118]

This very minimalist definition of attraction demonstrates perfectly two key elements of Diderot's methodology. First, it illuminates his concern for carefully untangling cause–effect relationships as they actually operate in nature, rather than in the imaginings of Man. Second, it is an example of his relative conservatism with regard to acknowledging the bounds of human knowledge. This spirit is revealed in an earlier statement as well: 'si l'attraction est une propriété générale de la matière, sentiment qui, pour ne rien dire de plus, est très probable, il est naturel de lui attribuer tous les effets qui lui sont analogues.'[119] Rather than pinning it down with a precise definition, Diderot ensures that his understanding of the issue remains philosophically defensible by not making any undue claims regarding the essence, nature, cause or even exact operation of attraction. Instead, he prefers simply to speak in terms of a force that exists in nature and is responsible in a very general way for certain observed effects.

The relevance of Diderot's acceptance of attraction goes well beyond the rejection of Descartes in favour of Newton; it was the tip of an iceberg. Just as Diderot posited that matter could not be viewed as homogeneous, so too he felt that admitting the force of attraction alone was insufficient to account for all observed phenomena. For him the acknowledgement of attraction was not the solution to physical problems, but only the beginning of an entirely new mode of understanding reality. It was here that the content of Diderot's physics pushed and eventually went beyond the boundaries of the Newtonian conception, just as his methodology had done vis à vis Bacon.

Diderot's physics is characterised, to a large degree, by an interest in what many mechanists of the day referred to as 'occult qualities'. This is not to say that Diderot accepted these qualities, much less that he was willing to employ them as an explanations any time there were unex-

118. *Sur la cohésion*, p.12.
119. *Sur la cohésion*, pp.9-10.

plained phenomena. Diderot was much too circumspect for that. This is evidenced in his attack on Maupertuis in *De la nature*, and the latter's unsound ascription to matter of the qualities of 'desire, aversion and memory'.[120] Nevertheless, Diderot never underestimated the wonderousness of nature, or perhaps better, never overestimated Man's knowledge of it. He was, in principle, always willing to acknowledge the existence of multiple, non-mechanical (as in non-contact action) forces should they be demonstrably in accordance with phenomena.

Another way of expressing this would be to say that a key aspect of Diderot's thought was an increasing awareness as to the importance of force, not simply as a concept, but as one of the key ontological elements of any physical explanation of reality. It might be said that Diderot came to see force as more crucial to considerations of reality than extension itself; he came to see matter more and more in terms of energy. In this respect, more than any other, we see the influence of Leibnizian thought on Diderot.[121] Diderot perceived that matter was not only acted upon by external forces or contact with external bodies, but that it was active in itself. He regarded force as being inherent to matter not simply in the sense that matter possesses force, but in that the two are indistinguishable. Diderot maintained that it was pointless folly, even in thought, to try to separate one from the other. What constituted an object, materially, was no more important than the manner in which it behaved relative to other objects by virtue of its inherent properties. Such a vision is made very apparent in pieces such as *Réfutation d'Hemsterhuis*, a work devoted primarily to combatting the mind–body dualism of what is now a relatively little known Dutch tract.[122] Here Diderot's thoughts on the issue of matter are laid out as concisely as anywhere within his canon of work. He proclaims that there is no such thing as 'absolute' rest,[123] that movement does not equal 'translation' (as in the translation of forces) and that the quantity of movement is not equal to (or entirely representative of) its nature.[124] Thus Diderot would eventually argue that the identity of a thing was constituted not so much by its material as by its continuity as an event. In this respect Diderot's consideration of matter was now entirely removed from the Cartesian conception of matter as inert.

All of this ties in neatly with another crucial element of Diderot's vision

120. *Pensées sur l'interprétation de la nature*, p.755-60.

121. See Cassirer's *Philosophy of the Enlightenment*, p.34-35. On the general influence of Leibniz on eighteenth-century French thought, see William Barber, *Leibniz in France, from Arnauld to Voltaire: a study in French reactions to Leibnizianism, 1670-1760* (Oxford, 1955).

122. Diderot and François Hemsterhuis (1721-1790) were actually acquainted with one another, having met in both Paris and the Hague. Georges May has assembled a useful text that provides the conflicting opinions of both men in one volume. See François Hemsterhuis, *Lettre sur l'homme et ses rapports, avec le commentaire inédit de Diderot*, ed. Georges May (New Haven, CT, 1964).

123. *Réfutation d'Hemsterhuis*, *OC*, vol.11, p.24.

124. *Réfutation d'Hemsterhuis*, p.26.

of the universe: its perpetual dynamism. Gone from Diderot's world-view is the Newtonian conception of a static universe. His emphasis is on change, movement and of a reality comprised of continually shifting forces. Everything, for Diderot, is in a state of continual creation and destruction. Early in his career, he advocated such a view in what might be described as a poetic fashion. His translation of Shaftesbury's *Merit and virtue*, for example, portrays a world in a state of continual flux, and the influence of this vision is witnessed in works such as *Lettre sur les aveugles*, with its casual assertion that change is one of the fundamental characteristics of the universe.[125] Later Diderot framed the idea in a more specific manner. In *Réfutation d'Hemsterhuis* he argued that the heterogeneity of matter was necessarily associated with a universe characterised by vicissitude.[126] *Matière et mouvement* is marked not only by its espousal of the heterogeneity of matter but its articulation of a world in which everything is in a state of 'translation' or 'in nisu'.[127]

We might even ascribe to Diderot here the development of an embryonic concept of entropy.[128] Granted, his idea is almost more poetic than in accordance with the laws of thermodynamics, but the fact remains that Diderot's universe is one in which the forces of creation are always counteracted by the forces of destruction, with the latter ultimately possessing something of an upper hand. Everything, according to Diderot, is eventually destroyed. Significantly the only exception to this principle is force itself. This is never dissipated but merely transferred to exert its influence somewhere else. It is in this sense, as well as that described above, that Diderot's philosophy eventually and ultimately puts more emphasis on force than on matter itself.

Thus a crucial part of Diderot's natural philosophical programme was an attempt to broaden the accepted definition of matter. Gone in Diderot's conception of the world was the traditional distinction between 'essential' and 'non-essential' properties. For Diderot, matter simply was. Extension was no more inherent to matter than any of its other observed properties; ultimately it was just an idea about matter, no more 'real' than notions such as that of number which was abstracted from the simple existence of bodies to form the basis of mathematics. Placed in historical context, Diderot can be viewed as dispensing entirely with Galileo's seminal distinction between primary and secondary qualities. His physics might also be perceived to have moved away from the viewpoint, shared by Galileo, Newton and Descartes, that understanding motion is the main problem of physics and that matter is simply assumed

125. *Lettre sur les aveugles*, p.199.

126. *Refutation d'Hemsterhuis*, p.30-31.

127. *Sur la matière et le mouvement*, *OC*, vol.9, p.609.

128. This has, indeed, been suggested by Geoffrey Bremner in *Order and chance: the pattern of Diderot's thought* (Cambridge, 1983), p.62.

to be inert and present only to be acted upon (be it by way of straightforward direct contact action or action at a distance). By conflating motion and matter as he did, Diderot placed the emphasis of experimental philosophy squarely upon matter and considerations regarding its nature, behaviour and potentially manifold properties and effects.

It must be remembered of course that in seeking to broaden the accepted definition of matter Diderot was not discarding mathematics, or mechanics, or the quantification and analysis of matter in motion. These pursuits remained for him quite valid, at least on a certain level. His insistence on re-evaluating contemporary notions of matter was therefore largely a semantic move, the ramifications of which could only become truly apparent further down the line of natural philosophical investigation. Diderot's crusade to broaden the definition of matter therefore had much to do with his methodology, and particularly his espousal of ignorance as a virtue. His stance may be summarised as proclaiming that while we do not know what matter is, we also do not know what it is not; therefore, if a theory that explains matter as inert fails to account for phenomena, we must be willing to admit that the observed effects are due to causes inherent in matter that are beyond our current knowledge (note how this relates to his aforementioned preoccupation with the notion of 'inappreciable physical causes'). In light of this, the goal of physics is not to reconstruct the world, but carefully to attempt to discover what the additional properties of matter might be. This is not unscientific, not necessarily purely descriptive, but simply cautious while remaining flexible and open-minded, as well as dealing with phenomena as observed effects rather than getting bogged down in a quagmire of disputes about cause.

One significant problem that remains is that against the backdrop of the position outlined above, the concept of physical law can easily become obscured. Jean Mayer, for instance, has suggested that overall Diderot remained resistant to the idea of physical laws.[129] Colm Kiernan and Wilda Anderson have likewise inferred that Diderot posits a universe in which there can be no such thing as fixed and immutable laws of nature.[130] The dynamic nature of the universe portrayed by Diderot would seem to make it difficult for him to posit any kind of order or permanent structure upon which Man can lay hold. This is a significant aspect of Diderot's world-view, and it also constitutes a compelling, almost poetic vision of what reality is like. But the assertion that Diderot did not believe in laws of physics needs careful consideration and qualification, not least because he himself did make reference to the concept of physical law. In *Matière et mouvement*, for example, he refers to the laws of

129. Mayer, *Diderot, homme de science*, p.200.

130. Colm Kiernan, 'Additional reflections on Diderot and science', in *Diderot studies* 14 (1971), p.136. Anderson, *Diderot's dream*, p.21.

movement not being erroneous[131] and in *Le Rêve de D'Alembert* to the general laws of physics.[132]

Of course, his notion of law was unorthodox. He defines a law of physics as the circumstances arising from the mutual 'rapport' between two or more various parts of matter.[133] It is, in other words, nothing more than the product of reciprocal action between two or more given elements of the universe.[134] Significantly the nature and terms of this reciprocal action were not necessarily seen by Diderot as continuously changing, at least in terms of form or structure. His universe may have resembled a cauldron, being one of perpetual movement and activity, but it was not entirely without regularity, or at least, not entirely without the possibility of being regarded as regular and uniform (even if, as 's Gravesande had asserted, this manner of regarding things was to some degree an assumption). On a general level certain repeated patterns of behaviour could be teased out of phenomena or perceived within it. This sense is bolstered on account of the fact that on many occasions Diderot referred to the idea of 'permanent phenomena'. A good example of this occurs in the article 'Encyclopédie', where he ascribes the phrase 'permanent phenomena' to the motion of celestial bodies.[135] Most strikingly, Diderot proclaimed in *L'Interprétation de la nature:* 'En ce sens général et abstrait, le système planétaire, l'univers n'est qu'un corps élastique: le chaos est une impossibilité; car il est un ordre essentiellement conséquent aux qualités primitives de la matière.'[136] It is clear then that Diderot does posit that the universe can be considered in some way uniform or noble, and that there is a sense in which the world can be known, in which reality can be approximated, and through which Man can gain some kind of handle upon it in order to make use of nature.

L'Interprétation de la nature indicates, for instance, that phenomena can and must be linked. The material gleaned from observation and experiment do not remain isolated, but are integrated with one another by means of reflection to form a system. This system may not possess absolute veracity, but this does not make it random or powerless. It is possible, within Diderot's conception, for ideas to be tested against experience; indeed, this is one of the cornerstones of his scientific methodology. The

131. *Matière et mouvement*, p.610.

132. *Le Rêve de D'Alembert*, p.103.

133. In using the word 'rapport' Diderot was deploying a term that had a number of interesting connotations in the context of eighteenth-century natural philosophy. Towards the beginning of the century the chemist E. F. Geoffroy had used the word to denote the affinity of one substance for another. Much later, Antoine-Laurent Jussieu was using the notion of 'rapports' to underlie his approach to natural history, which he conceived as a description of the relations between the observed phenomena of nature. See Emma Spary, *Utopia's garden: French natural history from old regime to revolution* (Chicago, 2000), ch.5.

134. *Sur la cohésion*, p.13.

135. 'Encyclopédie', p.420.

136. *Pensées sur l'interprétation de la nature*, p.741.

ideas themselves may only be approximations or relatively gross generalisations of that which actually exists, but strict empirical method and explanatory caution ensure that there is some relationship between these ideas about nature and nature itself.

Diderot's notion of the genius also has a role to play. By remaining flexible, adaptable and in tune with nature's demands, the ideal interpreter of nature gets as close as possible to the object of his study. Intuition is wedded to Man's rational faculties in the person of the genius, whose ideas necessarily develop in a fashion as dynamic as nature itself. By adhering to what must emerge as the primary moral of *Pensées sur l'interprétation de la nature* (that one should possess a system without being possessed by it)[137] the genius is free to alter his system in accordance with the demands of nature. This process of alteration is precisely what Diderot advocates in *Suite de l'Apologie de l'abbé de Prades*, where he argues that natural philosophy must undergo continual resystematisation.[138] The ever-changing nature of reality is dealt with then, at least to a certain degree, by Diderot's insistence that the content and even the methods of natural philosophy itself are ever changing and continually updated.

Thus, as we have seen, Diderot regards natural philosophical knowledge as fundamentally transient. But, while even the suggestion of this notion may have been entirely alien and abhorrent to many of Diderot's predecessors, it is not, in Diderot's eyes, a condemnation. Scientific knowledge may be tentative for Diderot, but he still regards it as knowledge. It must simply be seen as part of a process, rather than as a proclamation of absolute and eternal truth. Just like the world itself, Diderot conceives of natural philosophy as something that grows, develops and transforms itself. In this, he was not unlike Newton, whose fourth Rule of Reasoning in Philosophy stated: 'In experimental philosophy we are to look upon propositions inferred by general induction from phenomena as accurately or very nearly accurately true, notwithstanding any contrary hypotheses that may be imagined, till such time as other phenomena occur, by which they may be either made more accurate, or liable to exceptions.'[139]

Given these elements of his world-view, Diderot does have what might be described as some manner of 'faith' in science. The knowledge that it affords may be relative to those who have constructed it, but it is related to the world that exists. Furthermore it should be noted that given Diderot's epistemological sensitivities this subjective brand of truth is the only type possible and, for him, it only makes sense that all human

137. 'ayez un système, j'y consens; mais ne vous en laissez pas domineer', *Pensées sur l'interprétation de la nature*, p.731.

138. *Suite de l'Apologie de l'abbe de Prades*, p.635. See chapter 1, p.43 and n.110.

139. Newton, *Principia mathematica*, p.398.

knowledge should be related to humans themselves. Moreover this knowledge 'works', not just in terms of producing practical results but in terms of relating disparate phenomena to one another and offering some level of explanation for them. In other words, a model of reality can count as knowledge for Diderot, especially a model that has the capacity to become increasingly accurate and refined, both in terms of its technological applications and explanatory power (that is an increase in theoretical elegance or simplicity, the number of concepts explained with reference to certain laws or ideas etc.).

Finally one must consider that for Diderot there are different types of truth. This point is noted by Krakeur and Kruger, who note that in *Deux mémoires sur D'Alembert* Diderot makes a distinction between abstract science and mathematical physics. In the former instance, every problem is solved in the head of the mathematician, and all combinations are possible. In the second, phenomena are considered from the viewpoint of man, 'a being who passes like lightning and relates everything to his own duration'. Krakeur and Kruger hold that in maintaining this distinction, Diderot is positing two different and entirely contradictory types of truth.[140]

But another possibility is that these different types of truth are actually complementary. We might say that for Diderot there are different levels of truth, and that what is true on one level is not strictly true on another, although this does not completely falsify either. There are, in other words, a variety of valid ways of regarding the universe. Diderot expressed or hinted at this notion on several occasions throughout his career. In *Claude et Neron* he argues that the physicist can only formulate a theory regarding a machine which is itself regarded as a whole, but that the attainment of this theory becomes more removed (that is, less likely) as the machine becomes more complicated.[141] This would insinuate that there are various levels of theoretical complexity directly relative to the subject under investigation (which, in the broadest sense, is the universe). More tellingly, perhaps, the article 'Encyclopédie' notes that:

l'univers soit réel soit intelligible a une infinité de points de vue sous lesquels il peut être représenté, et le nombre des systèmes possibles de la connaissance humaine est aussi grand que celui de ces points de vue',[142] while *Pensées sur l'interprétation de la nature* notes that 'De même qu'en mathématiques, en examinant toutes les propriétés d'une courbe, on trouve que ce n'est que la même propriété présentée sous les faces différentes; dans la nature, on reconnaîtra, lorsque la physique expérimentale sera plus avancée, que tous les phénomènes, ou de la pesanteur, ou de l'élasticité, ou de l'attraction, ou du

140. Krakeur and Kruger, 'Mathematical writings', p.230.
141. *Claude et de Néron*, p.589.
142. 'Encyclopédie', *OC*, vol.2, p.406.

magnétisme, ou de l'électricité, ne sont que des faces différents de la même affection.[143]

Thus, Diderot's position is such that the proposition '1 + 1 = 2' is true in mathematics, but since mathematics is merely a system of generalisations abstracted from reality and not reality itself, the proposition is only true in a limited sense and has no ontological basis. Likewise Newton's inverse square law of gravitation is true in that it accurately accounts for the behaviour of physical bodies on a certain level of observation, but Diderot asserts that if a closer investigation of reality were possible, the situation would be infinitely more complex. Newton's formula works well enough, but it remains a generalisation; it is useful only when considering reality from a certain point, from a certain degree of removal. We might claim that Diderot has been vindicated here, since the development of quantum physics in the early twentieth century has indeed displaced Newton as a more accurate, intimate understanding of reality, yet upon a particular level of inspection, Newton's law of gravity and his other laws of motion still work – they still hold 'true'. The debate regarding the homogeneity of matter can be placed into this context as well. Rational mechanics, which treats matter as essentially homogeneous (simply considering factors related to extension, hardness, density and motion) does work on a certain level. Diderot asserts however that closer inspection of reality must reveal that matter is in fact heterogeneous, and that at what one might call a microscopic level the 'laws' of rational mechanics must ultimately fail. If reality is to be explained on this level, a new, more sophisticated system must be devised. This system, of course, is the discipline that has been under treatment here: experimental physics. But even experimental physics has its limits, a point at which it too becomes confounded. From there, one passes to an even more intimate level upon which to consider reality, a level dealing with the composition and internal constitution of matter. This level is chemistry, which shall be the subject of the next chapter.

143. *Pensée 45*, *Pensées sur l'interprétation de la nature*, *OC*, vol.2, p.752.

3. Chemistry

Sometime during the early 1750s the baron Paul Thiry d'Holbach introduced Diderot to the chemist Guilliame-François Rouelle (1703-1770). The reason for the meeting was so that Diderot might be persuaded to attend and compile a textbook on the chemistry course that was held annually by Rouelle at the Jardin du Roi.[1] It was a sensible piece of matchmaking on the Baron's part. Rouelle, though a brilliant lecturer and one of the most eminent chemists of his day, was not a writer. Diderot, though burdened at the time with the monumental task of editing the *Encyclopédie*, was a polymath who possessed not only the capacity to write, but to write technical prose. Diderot agreed to the scheme; perhaps as a favour to d'Holbach, perhaps out of enthusiasm to learn more about a scientific realm of which he knew very little.[2] In any case he began attending Rouelle's course in 1754 and went regularly for the next three years. Thus began what was to be for Diderot a lifelong affiliation with both Rouelle and his chemistry.

Despite this long-standing affiliation the role of chemistry in the development of Diderot's thought has been largely ignored. A number of studies devoted to Rouelle mention the fact that Diderot was among the chemist's most illustrious students, but the particulars of the manner in which the *philosophe* regarded and used the discipline fall outside the scope of these works.[3] Diderot scholars have offered little more in the way of illumination, making only vague and passing remarks about how chemistry might have contributed to his scientific world-view. Arthur Wilson mentions that the maturation of Diderot's thoughts on the ultimate nature of matter from the time of *Pensées sur l'interprétation de la nature* to that of *Matière et mouvement* may have had something to do with his attendance of Rouelle's lectures, but does not really pursue the issue any further.[4] Perhaps the fullest treatment of the issue is Jean Mayer's suggestion that Rouelle's chemistry course furnished Diderot with two seminal concepts: that of the transformation of matter and that of vital chemical energy, and that armed with these ideas he was able to deride

1. Mayer, *Diderot, homme de science*, p.42.

2. D'Holbach himself possessed a fairly impressive knowledge of chemistry, having translated a number of German mineralogical texts for the French market, and written several chemistry related articles for the *Encyclopédie*. For more on this aspect of his career, see Pierre Naville, *D'Holbach et la philosophie scientifique au XVIII siècle* (Paris, 1967), p.75-77.

3. The best overview of Rouelle's life, work, and significance is Rhoda Rappaport, 'G.-F. Rouelle: an eighteenth century chemist and teacher', in *Chymia*, vol.6 (1960), p.68-101. This article includes a bibliography on other studies of Rouelle.

4. Wilson, *Diderot*, p.580.

the mechanistic view of matter.[5] Simple statements such as this represent the extent to which Diderot's interest in and application of chemistry have been explored.

The primary aim of the current chapter is to redress this situation by offering a thorough examination of the manner in which G. F. Rouelle's chemistry contributed to Diderot's natural philosophical thought. At root it is a study in influence; a study of how certain ideas expressed by one of the mid eighteenth century's most renowned scientists were interpreted, assimilated and given renewed expression by one of the eighteenth century's most prolific philosophers. As such this chapter acquires a character rather different from any of the others in the present book. Because chemistry stands out as the only natural philosophical discipline in which Diderot's work and thought can be traced overwhelmingly to a single source, it shall be treated here with almost as much attention to the nature and content of that source as to Diderot's ideas themselves. The result is a chapter of two rather distinct halves. The first part deals with the chemistry course of Rouelle and examines the nature of the information with which Diderot was presented between the years 1754 and 1757. The second part of the chapter focuses on Diderot's rising awareness and appreciation of chemistry and comes to terms with the impact of chemistry on many of his later works. The latter task is of paramount importance, for it is only after achieving a proper understanding of how Diderot employed concepts inherited from Rouelle that much of his mature thought becomes truly intelligible and meaningful.

Part of the reason that the subject of Diderot's interest in chemistry has attracted so little attention over the years may be that his primary instructor has left what might be described as an ambiguous legacy. Historians of science are divided as to how novel, influential and important Rouelle's work and teachings are. Many accord him only a minor place in the history of chemistry, maintaining that he represented a school of thought which was antiquated even in its own time, and that his achievements are few and of little actual significance. More sympathetic commentators rate Rouelle very highly, and credit him with having exerted a marked influence on the development of eighteenth-century science by laying foundations without which the 'Chemical Revolution' of the 1770s and 1780s would have been impossible.[6] The argument here is

5. Diderot, *Eléments de physiologie*, ed. Jean Mayer (Paris, 1964), p.lvi.

6. This opinion was set forth by one of Rouelle's earliest biographers, Paul-Antoine Cap, and it has been maintained in more recent times by Rhoda Rappaport. For Cap's appraisal, see Paul-Antoine Cap, *Guilliame-François Rouelle* (Paris, 1842) p.30. Rappaport's analysis can be found in 'Rouelle: eighteenth century chemist', as well as in her doctoral thesis 'G.-F. Rouelle: His *cours de chimie* and their significance for eighteenth century chemistry', Cornell University, NY, 1958. On the nature, process and meaning of the 'chemical revolution' itself, see H. Gilman McCann, *Chemistry transformed: the paradigmatic shift from phlogiston to oxygen* (Norwood, NJ, 1978); Frederic Lawrence Holmes, *Eighteenth-century chemistry as an investigative enterprise* (Berkeley, CA, 1989), ch.5; Maurice Crosland, 'Chemistry and the

that while the extent of his discoveries and of his contributions to theory may have been modest, he did manage to synthesise a number of crucial trends, conducted an extraordinary amount of research, participated in the process of rationalising chemical nomenclature by trying to eliminate redundancies, and of course almost single-handedly educated an entire generation of French chemists which included Gabriel François Venel (1723-1775), Pierre Joseph Macquer (1718-1784) and most famously Antoine Laurent Lavoisier (1743-1794).[7]

Irrespective of historiographical stance, there are a number of things that can be affirmed with regard to Rouelle's career. First he was not, as has been commonly asserted, the first to bring Stahlian chemistry into France.[8] That distinction belongs to Étienne-François Geoffroy (1672-1731), a *physicien* who belonged to a slightly earlier generation than Rouelle.[9] Nevertheless Rouelle did play an important role in propagating the chemistry of Stahl, albeit in modified form.[10] Perhaps the most crucial aspect of Stahl's work, and certainly one that was to maintain a firm hold over the imaginations of many French chemists for years, was its espousal of the phlogiston theory. This theory posited an inflammable principle (phlogiston) which was separated from bodies subjected to heat, thus accounting both for the visible flame and for the fact that substances, particularly metals, were reduced in volume after such a process. Not everyone in France subscribed to the theory; Buffon, for example, wrote it off as purely speculative and unsubstantiated nonsense.[11] Nevertheless, the phlogiston theory did play an important role in guiding the course of eighteenth-century French (and English) chemical thought. Even

chemical revolution', in *The Ferment of knowledge*, ed. G. S. Rousseau and Roy Porter (London, 1980), p.389-416; Evan Melhado, 'Chemistry, physics, and the chemical revolution', in *Isis*, vol.76, 1985, p.195-211; and Alistair Duncan, *Laws and order in eighteenth-century chemistry* (Oxford, 1996).

7. Rappaport, 'Rouelle: eighteenth century chemist', p.77. On Lavoisier, see Henry Guerlac, *Lavoisier – the crucial year: the background and origin of his first experiments on combustion in 1772* (New York, 1961); and Frederic Lawrence Holmes, *Antoine Lavoisier – the next crucial year* (Princeton, NY, 1997). Another of the most noteworthy attendees was Rousseau, who followed the course at the same time as Diderot. See Diderot, *Eléments de physiologie*, ed. Paulo Quintili (Paris, 2002), p.16.

8. Georg Ernst Stahl (1660-1734) was a professor of medicine at the University of Halle. He was adept at the then relatively new science of medical chemistry, and his work would have important repercussions on eighteenth-century thought in both disciplines. A classic overview of his chemical theory is Hélène Metzger's, *Newton, Stahl, Boerhaave et la doctrine chimique* (Paris, 1974). Regarding his medical interests and views pertaining to the nature of life, see chapters 4 and 5 of this book.

9. J. R. Partington, *A History of chemistry*, vol.3 (London, 1962), p.51.

10. Rhoda Rappaport, 'Rouelle and Stahl – the phlogistic revolution in France', in *Chymia*, vol.7 (1961), p.73-102.

11. The first volume of Buffon's *Supplement* to *Histoire naturelle* (1774) launched a veritable campaign against the notion of phlogiston. See Bernadette Bensande-Vincent and Isabelle Stengers, *A History of chemistry*, translated by Deborah van Dam (Cambridge, MA, 1996), p.87.

Lavoisier, who eventually debunked the theory once and for all, was not entirely free of its influence. His notion of caloric, which addresses the same basic issues (heat, fire and the reason that substances burn) seems to have been conditioned to at least some degree by the great debate about the existence of phlogiston.[12] Thus, by advocating Stahl's phlogiston idea, Rouelle promoted an element without which it is virtually impossible to talk about pre-nineteenth-century chemistry.

Although the epistemological status of the phlogiston theory may seem dubious and abstract to modern readers, Rouelle was actually a very prominent empiricist who devoted much more energy to the practice of chemistry than to the development of theory. Maintaining the motto 'Nihil est in intellectu quod non prius fuerit in sensu',[13] he was one of the mid eighteenth century's greatest proponents of experimental science, advocating the need to develop and perfect instrumentation, and to exhibit great caution, precision and finesse in the conduct of experiments. Always vociferous about staying close to 'facts', Rouelle felt that the only sound theory was one based entirely on empirically obtained results. For him, ideas were relegated to the role of merely forming links between experiments.[14] Getting carried away with concepts he regarded as one of the cardinal sins of scientific practice.

A career as a devoted experimentalist bore significant fruit for Rouelle. His best-known discoveries are those pertaining to salt, which is the subject of a number of memoirs he read to the Académie des sciences in the 1740s and 1750s.[15] Taking as his starting point the classification of salts according to crystalline structure, Rouelle proceeded to unravel the nature of their composition by referring to the interplay between acids and bases. Thus Rouelle established three basic categories of salt: those in which acid is superabundant, those in which a base is superabundant, and those in which they are in equilibrium.[16] This body of work impelled the study of salts to go beyond mere form and into more complex realms of analysis. It was also connected to important advancements in the notion of saturation points and to ground-breaking discoveries regarding

12. For a succinct overview of the impact of the phlogiston theory on Lavoisier's thought, see Paul Strathern, *Mendeleyev's dream: the quest for the elements* (London, 2001), ch.9 and 10. On phlogiston in general, see John Henry White, *The History of the phlogiston theory* (London, 1932).

13. Rappaport, 'Rouelle: eighteenth century chemist', p.75.

14. Rappaport, 'Rouelle: eighteenth century chemist', p.75.

15. See: *Mémoires de l'Académie des sciences*, 1744, 1745 and 1754. The subject of salts attracted a great deal of attention from eighteenth-century chemists, primarily on account of the fact that it provided a relatively neat opportunity to give the discipline some sort of rational structure, and because it dealt directly with the intriguing question of substances (in this case acids and alkalis) that form compounds in which the properties of the constituents seem lost, or at least inactive. For more on these issues consult Bensaude-Vincent and Stengers, *History of chemistry*, p.53.

16. Cap, *Rouelle*, p.20.

the properties and behaviour of acids and bases. Most famous with respect to the latter area of research was Rouelle's demonstration that essential oils could be ignited with the acid 'spirit of nitre' (nitric acid).

Rouelle also worked a great deal on the vegetable and animal kingdoms of matter, his labours enhancing chemical understanding of both these realms. Indeed one of Rouelle's first biographers argued that he might be regarded as the father of vegetable chemistry.[17] While this claim may be rather grandiose, it is certain that Rouelle's analyses of vegetable and animal matter assisted the development of chemistry in two ways. First, in distinguishing between the diverse materials of his many analyses, Rouelle added immeasurably to empirical knowledge of the parts that constitute various organic bodies. Going far beyond the work of his predecessors Hermann Boerhaave (1688-1738) and E.-F. Geoffroy, Rouelle deeply enhanced knowledge of substances such as blood, urine, milk and bile.[18] Second, desire to obtain more intimate knowledge of these substances led Rouelle to develop or perfect a number of new instruments and techniques. An example of this is his work on methods of distillation, which was to be an indispensable tool not only for himself but for future generations of chemists.[19]

Despite this vast amount of work, the fact remains that Rouelle's most significant contribution was in his role as an educator. His course, held for six to eight months each year from 1742 to 1768, was responsible for raising the profile of chemistry in France not only within scientific circles, but amongst the general public.[20] Testament to this is the sheer popularity of his lectures. They were held at the Jardin du Roi, which by this time had become a well-established centre for demonstrations not only in chemistry but in disciplines such as botany and anatomy as well.[21] Significantly, admittance to these sessions was completely free. Accessibility, however, was certainly not the only reason for the success of Rouelle's course. Evidence of this is that many of those to attend were members of Paris' most elite social circles. Many of these patrons could have easily afforded an entrance fee, and in fact some even did pay to attend private demonstrations in Rouelle's laboratory. The factor that is likely to have drawn 'pupils' from this social set is that, at least for a time, the lectures of Rouelle were regarded as fashionable, and were one of

17. Cap, *Rouelle*, p.18.

18. Cap, *Rouelle*, p.12. For more on Boerhaave (who, like Stahl, was a doctor of medicine who took a profound interest in both the theory and practice of chemistry) see Metzger, *La doctrine chimique*, as well as Gerrit Arie Lindeboom, *Hermann Boerhaave: the man and his work* (London, 1968).

19. Charles Henry, 'Introduction à la chymie; Manuscrit inédit de Diderot', in *Revue scientifique*, 2e semestre 1884 (3e série), no.4 (26 Juillet 1884), p.108.

20. Henry, 'Introduction a la chymie', p.108.

21. On the pedagogical role of the Jardin du Roi, see Spary, *Utopia's garden*, and J. P. Constant, *L'Enseignement de la chimie au Jardin Royal des Plantes de Paris* (Cahors, 1952).

Paris' premier locations at which 'to be seen'.[22] Thus, in the words of Diderot, 'the quarter [of the city] with the meanest populace became the meeting place for all classes, not excluding the children of nobles who wanted to learn'.[23]

There are two other elements of Rouelle's course which undoubtedly drew a great many people. First, his demonstrations were often spectacular, frequently featuring violent and dangerous explosions (some of these intentional, some not!). Perhaps even more important was the rather striking nature of the man himself. By all accounts, including those of Diderot, Rouelle was an extremely charismatic and exuberant man who taught in a dramatic and compelling manner.[24] As one former student proclaimed:

> When Rouelle spoke, he inspired, he overwhelmed; he made me love an art about which I had not the least notion; Rouelle enlightened me, converted me; it is he who made me a supporter of that science [chemistry] which should regenerate all the arts, one after the other [...]; without Rouelle, I would not have known how to look above the mortar of the apothecary.[25]

Diderot attended these lectures from 1754 to 1757. Unfortunately the notes that he took while in attendance have long since been lost. What is extant are manuscript copies of the textbook that Diderot compiled on the basis of these notes. One of these is in Bordeaux (Bm Bordeaux, Ms. 564), the other in Paris (B.N., n.a.fr. 4043-4044). While the Bordeaux manuscript is the only one to bear Diderot's signature, both versions are unquestionably in his hand and are identical in all respects but for the fact that the Bordeaux copy includes a brief essay on the history of chemistry.[26] It is from one of these manuscripts that we must glean the content and nature of the chemistry presented to Diderot in the mid 1750s.

There is a paradox, of course, in employing a text prepared by Diderot himself as the basis for understanding something that is said to have influenced much of his later work. Any manuscript based on Diderot's notes is not going to be a simple, objective portrayal of Rouelle's chemistry. In the very act of being recorded by Diderot, the content of Rouelle's lectures will necessarily have been, to one degree or another, interpreted. This fact must be borne in mind when using the text as the source for outlining what Diderot learned from Rouelle. There are

22. Cap, 'Rouelle', p.19.

23. Rappaport, 'Rouelle: eighteenth century chemist', p.72.

24. Diderot's obituary of Rouelle (who died in 1770) provides a lively and intimate overview of Rouelle's often eccentric personality. See *Note historique sur Michel Vanloo, peintre, et le chimiste Rouelle*, *OC*, vol.9, p.599-603.

25. Louis-Sebastien Mercier, *Tableau de Paris*, 12 vols (Amsterdam, [s.n.], 1782-88), vol.11, p.178-79.

26. See J. Mayer, 'Cours de chimie de Rouelle', in *Diderot: œuvres complètes*, vol.9, (Paris, 1981), p.180.

however a number of compelling reasons for using it nonetheless. First, although more than twenty manuscripts describing Rouelle's lectures are extant, it is Diderot's work that provides the clearest, most orderly, most comprehensive account. Indeed, it has been argued that the primary reason Diderot was approached to compile a textbook was Rouelle's dissatisfaction with the student notes on his course that were circulating at the time.[27] Thus there is a good chance that Diderot's version is not only the most accessible and the easiest to read, but also the one that most pleased Rouelle. Second, comparison of Diderot's manuscript with some of these other, rather haphazard editions of Rouelle's course reveals that its basic content and representation of most seminal ideas is not wayward.[28] The most significant difference between Diderot's text and any of the others is primarily in terms of eloquence, thoroughness and sense of structure. Finally, the content of Rouelle's lectures inevitably changed over time. When he began teaching at the Jardin du Roi, Rouelle had not yet arrived at many of the ideas that he would later propound. In light of this, it is most sensible to base an overview of what Diderot learned on information that was recorded at the time of his attendance. Nothing could do this better, of course, than his own account.

Attention may now be turned to the text itself, the full title of which is *Cours de chimie, ou Leçons de Mr. Rouelle, recueilles pendant les années 1754, et 1755; Rédigées en 1756; Revues et corrigées en 1757 et 1758.* Totalling 694 pages (in folio), it is divided into two volumes. The first volume contains an introduction and chapters on both the 'règne vegetal' and the 'règne animal'. Volume 2 is devoted almost exclusively to the 'règne mineral', although it does incorporate a brief concluding passage on alchemy.

The course begins with an exposition on the nature of chemistry. Describing the discipline as 'un art physique', it argues that with the proper application of operations and instruments, chemistry leads to certain knowledge of the constitution of physical bodies.[29] It does this by separating substances into their constituent parts (or at least into the simplest form that can be obtained) in order to understand their composition. It is also concerned with taking these parts and either recombining them, or mixing them with different parts to create entirely

27. See Mayer, 'Cours de chimie de Rouelle', p.179.

28. An example of such a manuscript is B.N., n.a.fr. 4045, *Cours de chimie par M. Rouelle*, 1751. This piece typifies those notes on Rouelle's course dated prior to 1754 in that it tends to be rather confused and awkwardly expressed. Perhaps more importantly, many of these early compilations are incomplete. There are also many later editions of the course, most of them c.1760. Not only do most of these texts conform to that assembled by Diderot, but many of them seem to be based primarily on his notes. Two of these, at the Wellcome Trust Library in London (Manuscripts 4281/2 and 4283), explicitly acknowledge this debt. For more on these issues, see Jean Jacques, 'Le Cours de chimie de G.-F. Rouelle recueilli par Diderot', in *Revue d'histoire des sciences*, vol.38:1 (Jan-Mar 1985), p.43-54.

29. *Cours de chimie*, p.1. I am using the Paris manuscript (B.N., n.a.fr. 4043-4044).

new substances. Thus the basis of chemistry is really nothing other than analysis and synthesis, or the decomposition and recomposition of matter. According to Rouelle, this process of taking things apart and putting them back together again makes chemistry a universal science; its method is something that can and should be tried on everything. Despite this universality, Rouelle was at pains to point out that chemistry is occupied with particular and singular knowledge of bodies, not with sweeping generalities. It is a discipline that searches not for vain 'raisonnemens', but for facts.[30] Chemistry is put forward here as the pre-eminent experimental discipline.

The introduction to the course also discusses the practical value of chemistry. Although this seems a far from original concept, it is one that cannot be overemphasised. Men had tried since antiquity and throughout the Middle-Ages to make useful application of what is now called chemistry, but in the middle of the eighteenth century appreciation for the utility of the discipline was rising to new heights. Chemistry was slowly emerging from its role as a mere adjunct to the fields of mining, metallurgy and medicine; it was establishing itself as an autonomous discipline and one that was applicable to all of the arts. The Rouelle manuscript makes this point forcefully, and in so doing serves as a valuable marker in tracing the rise of chemistry as a proper, and valuable, branch of science. Having already argued that the method of chemistry is applicable to all types of matter, Rouelle notes its importance to arts such as glass blowing, metallurgy and warfare. The main focus, however, is on the indispensability of chemistry to medicine. Rouelle argues that chemistry is not only a very effective tributary to the art of healing, but that it explains phenomena without which medicine would be unintelligible. Only through chemistry, the manuscript tells us, can one know anything about changes of state, liqueurs, putrefaction, gangrene, blood, etc.[31] Rouelle, it seems, is trying to establish chemistry as a body of knowledge that not only serves pharmacy, but underlies physiology as well.

This argument in favour of the utility of chemistry is not just restricted to the introduction, but is reinforced throughout. Almost every operation performed under the section 'règne végétal' ends with some manner of discussion regarding the medicinal properties of the substance or substances obtained. Likewise, most of the procedures of the 'règne animal' reveal the composition and properties of various bodily humours. Each substance discussed in the 'règne minéral' section is treated first with regard to the manner in which it is assayed and mined. The course also includes detailed discussion, sometimes many pages in length, on the

30. *Cours de chimie*, p.2.
31. *Cours de chimie*, p.2.

chemical nature and process of making various items such as porcelain, wine, beer and soap.

After outlining the basic nature of chemistry, Rouelle's course begins in earnest with an explanation of the most fundamental concepts and terms. The first of these is 'principes', or principles. 'Principes' correspond to what we would today understand as chemical elements, in that they are the building blocks of which everything is composed. To draw a closer resemblance, however, between Rouelle's 'principes' and the elements of modern chemistry would be anachronistic and misleading. First, Rouelle only posited four principles: the classical earth, air, fire and water. The manuscript tells us that he contemplated adding a fifth, the mercurial principle of Stahl's mentor Johann Becher (1635-1682),[32] but as there are few references in the course to this fifth principle, it is safe to assume that Rouelle did not regard it as especially significant. In any case, by maintaining these basic elements as the foundations of all matter, Rouelle was following a tradition that had been established since ancient times.

More critical is the fact that for Rouelle principles were never to be found pure in nature; indeed they were not even observable and were therefore lacking experimental verification. Despite this, he claimed that they are indivisible, homogeneous and more or less mobile depending on their particular nature, figure and mass.[33] It may seem contradictory, of course, to speak of the specific attributes of bodies that are by definition totally insensible. It must be noted, however, that although Rouelle spoke of principles as having a specific figure, mass, even volume, he never tried to specify what these were, and acknowledged that it would be impossible to try to find out. It is simply assumed that because principles are the constitutive elements of matter, that they must possess the defining characteristics of matter itself.

Of course the modern scientist may still wonder, given the entirely speculative nature of principles, why anyone would bother with them at all. To this, two responses must be given. First, the need had not yet arisen within chemistry to try to define the most basic building block possible in terms of the simplest substance that may be experimentally discovered. That was one of Lavoisier's primary contributions and although Rouelle may have been moving in that direction, it was not essential, given the kind of chemistry that he was trying to do, to posit such a definition. Second, as abstract as the notion may seem, principles were necessary to account for the observed properties of objects. They were regarded as the bearers of quality. For example, it is easy enough to say that substance A is a mixture of substances B and C, but why is it

32. *Cours de chimie*, p.6. Becher's own theories are propounded in his *Physica subterranea*, which appeared in 1667. More on his life and work can be found in chapter 4 of Bernard Jaffe, *Crucibles: the story of chemistry from ancient alchemy to nuclear fission* (New York, 1998).

33. Cours de chimie, p.5.

transparent, or hard, or susceptible to combustion? It is here that principles are invoked, almost as metaphors, to explain why substances have the characteristics they do. Thus fire lends things subtlety, air elasticity, and earth solidity, to name just a few of the qualities for which principles are ultimately responsible.

Dealing with Rouelle's description of precisely how principles become joined to form sensible bodies is just as tricky, from a methodological point of view, as the concept of the principle itself. In a frustratingly brief passage, he mentions that they are united with one another not through penetration, but through 'juxtaposition'.[34] Rouelle does not attempt to provide a precise or detailed explanation of what he means by this, but it can at least be surmised that in rejecting penetration, he is declaring his opposition to mechanical theories about the manner in which matter is held together. Unfortunately, the potentially intriguing concept of 'juxtaposition' (which seems to be suggestive of what contemporaries might have referred to as 'occult causes') is left unpacked, and the course briskly moves on to discuss the various orders of physically sensible matter: 'agrégés', 'mixtes', 'composés' and 'surcomposés'. The simplest kinds of observable bodies are aggregates (homogeneous bodies consisting of a single principle) and 'mixtes' (heterogeneous bodies composed of two or more principles).[35] Rouelle says that they are held together by 'cohesion', which is another force upon which he fails to elaborate.[36] He does say, however, that the union which forms aggregates and 'mixtes' is very strong and difficult to break apart.[37] This is not so for 'composés', which are formed by the combination of aggregates and/or 'mixtes' with one another, and as such are more tenuously bound. 'Composés', in turn, unite with one another (or indeed to aggregates or 'mixtes') to create 'surcomposés', which are formed from the amalgamation of so many elements that they come apart relatively easily.[38] Thus, Rouelle develops

34. *Cours de chimie*, p.6.

35. For Stahl, aggregates were distinguished from mixtures in that they were only held together by simple, mechanical bonds, whereas mixtures were conjoined in a truly chemical fashion (see Bensaude-Vincent and Stengers, *History of chemistry*, p.58-59). Rouelle does not comment on Stahl's doctrine here, but his suggestion that principles are never held together by interpenetration, but rather through 'juxtaposition' seems to indicate that he felt all orders of combination to be chemical, without any room for the mechanical whatsoever. For Rouelle, even aggregates are bound together in a manner far more intimate than that of simple, physical attachment.

36. In the wake of Newton, the possibility that the smallest particles of matter might be held together by gravitational attraction (or some form of it) was extensively considered and debated, especially in England. In this connection, it is worth noting that either there was more on cohesion in Rouelle's lectures than comes across in Diderot's notes, or that Diderot was influenced in his thoughts on the matter by some other source(s). This can be surmised from the fact that Diderot gives the concept of cohesion fairly rigorous and thoughtful treatment in works such as *Sur la cohésion* (see previous chapters).

37. *Cours de chimie*, p.6.

38. *Cours de chimie*, p.7.

something of a scale or ladder of composition that ranges from the simplest substances (which are tightly bundled together and relatively few in number) to the most complex (which are nothing more than combinations of the simpler forms).

The final section of the introduction describes the chief instruments of chemistry, four of them natural, two of them artificial. The natural instruments are nothing other than the principles themselves (in their observable, that is, aggregated state, of course[39]). Of the four main principles the two most important here are fire and water, which correspond to the 'dry way' and the 'wet way' of chemical analysis. Fire is described as the principal motor, the 'first cause of almost all of the effects of Nature'.[40] It agitates aggregated bodies, separating molecules from one another in the process. As such, it is tremendously useful in the decomposition of substances. Water, apart from supposedly entering into the composition of nearly every body, can be used as a solvent and is therefore another of the chemist's indispensable tools. In this we see an example of why Rouelle's chemical system has earned the epithet 'the element-instrument theory' of chemistry.[41]

The two artificial instruments of chemistry are classified as 'vaisseaux' and 'menstrues'. The first of these is relatively self-explanatory, although Rouelle does spend a good deal of time describing how his furnaces, containers and mechanisms of distillation work. Perhaps most interesting is Rouelle's claim to have perfected Stephen Hales' (1677-1761) machine for measuring, through the displacement of water, the amount of air disengaged from decomposing bodies.[42] He also lists a number of other processes critical to the practice of chemistry, such as filtration, clarification, trituration and evaporation, but notes that these should properly

39. It is also important to distinguish between principles in their true form (as unobservable and essentially metaphysical, descriptive entities) and principles in their aggregated form. The latter, while being the only physical manifestations of principles, do not necessarily share all of the properties of the former. For example, Rouelle believes that water as a principle has nothing to do with fluidity, but that in a state of aggregation, it is fluid, insipid, transparent and freezable. Even more interesting is his conviction that the fluidity of water in its aggregated state is actually due to its inclusion of a minute element of fire.

40. *Cours de chimie*, p.11.

41. Rappaport, '*Rouelle and Stahl*', p.93-95. In a refinement of the ideas of Stahl, Rouelle held not only that principles are the building blocks which combine to create the various tangible substances of everyday life, but that at the same time they are the very agent of chemical reactions as well. This is why, according to Rouelle, principles could not be isolated; a principle is only separated from one mixte on account of entering into another (Bensaude-Vincent and Stengers, *History of chemistry*, p.62.).

42. *Cours de chimie*, p.42. Following Hales and recognising not only that air is a principle but that it has an important chemical role was another respect in which Rouelle diverged from Stahl. See Bensande-Vincent and Stengers, *History of chemistry*, p.62. On life and work of Hales himself, see D. G. C. Allan and R. E. Schofield, *Stephen Hales, scientist and philanthropist* (London, 1980).

be referred to as preparations rather than operations, since they are not responsible for decomposing bodies or for composing new ones.[43]

The issue of 'menstrues' is less technical, but more complex. This is because it deals with dissolution, and attempts to account for why, in a solution of water, acid, base or any other fluid, certain bodies have a tendency to leave their current partnership and enter into a new one with the solvent. Rouelle's discussion of this phenomenon can be divided into two sections. The first deals with the simple fact that part of a dissolved body can actually unite to the solvent. Rouelle maintains an agnostic position regarding the cause of such unions, saying that whatever it may be, it is totally unknown to us. Nevertheless he does say that the tendency of certain molecules to combine with one another must have a cause, for without one the molecules set free in a dissolution would be entirely aimless and the liquid opaque. He also states that this cause must be a virtue inherent in the bodies themselves.[44]

The second section on 'menstrues' deals with chemical affinity, or the fact that certain bodies tend to combine with one substance rather than another. Rouelle's discussion of chemical affinity, or 'rapport' as he seems to prefer, is only slightly clearer than his discussion regarding the nature of principles. The lack of clarity here is once again largely due to lack of elaboration. What is stated is that the attraction of certain bodies to one another is not due to the law of the square of distances as in Newtonian (gravitational) attraction, but to that of the homogeneity of surfaces.[45] Presumably, Rouelle is referring here to Stahl's 'latus' theory, which asserted that substances are attracted to one another on the basis of the similarity of the principles which constitute them. He goes on to mention that the 'cohésion' of parts of the body dissolved to those of the solvent is the cause of their suspension and the transparence of the solution. This seems to be a mere repetition of his views on chemical attraction. Concluding the discussion, Rouelle states that this cause (as imperfect as our knowledge of it may be), is a better explanation of phenomenon

43. *Cours de chimie*, p.45.

44. *Cours de chimie*, p.33.

45. *Cours de chimie*, p.34-35. In denying the ability of Newtonian attraction to account for chemical affinity, Rouelle distinguished himself from a number of those who would come after. More will be said on this in due course, when we discuss Diderot's views on the matter. Another respect in which reference to Newton here is of interest, is that it was the great Englishman's *Opticks* which in many ways became a departure point for discussion of chemical affinity in the eighteenth century. In 'Query 31' of the *Opticks* Newton attempted to come to grips with the question of why certain bodies prefer combination with one substance as opposed to another, and he seems to approve of the notion that particles of matter can indeed possess qualities (or 'active principles') which account for such phenomena (Newton, *Opticks: or treatise of the reflections, refractions, inflections and colours of light* (New York, 1952), pp.375-76). For more on this aspect of his work and influence, see Bensande-Vincent and Stengers, *History of chemistry*, p.48-50; Duncan, *Laws and order*, p.42-51 and 72.

than any of the systems dealing with 'points of menstrues' and 'pores of bodies'. Rouelle is here quite clearly showing disdain for the mechanistic systems of Boyle and many of the other English chemists of the late seventeenth and early eighteenth centuries.[46]

What can be concluded from this rather hasty discussion of one of eighteenth-century chemistry's most crucial concepts? It is clear that Rouelle was not willing to accept any of the mechanical explanations for chemical affinity (or the manner in which matter coheres at all), nor was he interested in explanations that depended upon some form of gravitational attraction. He seems instead to have been thinking along the lines of what may be termed purely chemical causes as explanations for affinity and for the binding of particles to one another. The existence and operation of such causes would certainly have been an issue worth explicating (if only on account of their contentiousness), but as we have seen, this was not forthcoming from Rouelle. A very real possibility as to why this may have been the case is that the entire issue of underlying causes was actually not very important to him. Why, from a metaphysical point of view, a particular substance is drawn to one rather than another is of less interest to him than the simple fact that it happens and can be observed. Although there are a number of references throughout his course to substances combining on account of this or that 'latus', none of his arguments or demonstrations ever depend on the theoretical validity of this notion. He seems to be more or less content with performing his experiments, observing which substances have a 'rapport' for one another, and cataloguing the results. According to Rouelle, one can never really anticipate in advance what substances might have a preference for one another, therefore the chemist needs to discover these relationships by trial and error anyway. In taking this kind of stance, Rouelle was very much in the tradition of mid-eighteenth-century French chemistry, which tended to shun attraction as the motivating force behind chemical combination and accept a rather more empirical view, simply defining chemical affinity as 'the tendency of things to combine'.[47]

It is here that the introduction ends. The remainder of the course is devoted to the procedures, or experiments themselves. These procedures, which represent the vast bulk of the course, are divided into three categories; those of the 'règne végétal', those of the 'règne animal' and those of the 'règne minéral'. It may have been expected, given Rouelle's

46. *Cours de chimie*, p.35. On Boyle's corpuscularian philosophy and its ramifications regarding chemistry, see Peter Alexander, *Ideas, qualities and corpuscles : Locke and Boyle on the external world* (Cambridge, 1985) and also Marie Boas Hall, *Robert Boyle and seventeenth-century chemistry* (Cambridge, 1968).

47. Duncan, *Laws and order*, p.98. The framework for this tradition was sketched out, to a large degree, by Geoffroy, who drew up the first table of chemical affinities; working them out on the basis of displacement reactions (Bensande-Vincent and Stengers, *History of chemistry*, p.67).

training as a pharmacist and his predilection for vegetable and animal chemistry, that these two areas of inquiry would dominate the course, but this is not the case. In fact, the section 'règne minéral' comprises approximately 60 per cent of the manuscript. It includes 159 procedures over 375 pages, versus 56 over 190 for the 'règne végétal' and 11 over 41 for the 'règne animal'. This can be taken as very indicative of the state of mid-eighteenth-century chemistry. Although analysis of organic substances was beginning to develop, in part due to the efforts of Rouelle himself, knowledge of the organic kingdoms still lagged far behind that of the mineral world. This was largely due to the role of mining and metallurgy in stimulating the development of chemistry as early as the seventeenth century. By contrast medicine, which eventually played a role in prompting the growth of vegetable chemistry, had long been reliant on botany. Even in the 1750s, the notion that chemical study of the 'règne végétal' might be able to contribute more to medicine than botany was still a relatively contentious concept. Thus, Rouelle devotes an entire discussion to defending the efficacy of chemistry with regard to medical matters. He argues that chemistry is superior to botany, and that it warrants greater attention from the medical establishment because it provides more nuance; it goes beyond mere form and seeks to establish the true nature of things.[48] Animal chemistry, of course, was even less developed. The body of knowledge comprising the 'règne animal' was in a state of infancy and this is reflected by the fact that, despite Rouelle's interest in the area, there is comparatively little he can say about it.

The order in which Rouelle chose to present the various kingdoms of matter is also revealing. He commences with the procedures pertaining to the 'règne végétal' and the 'règne animal' respectively, despite the fact that these are the most complex. The reason that they are more complex is that they are more composed and, as the manuscript tells us, 'the order of analysis begins with the bodies that are most composed, because these are the easiest to separate, and because knowledge of them naturally leads to knowledge of simpler substances which are more difficult to decompose'.[49] Of course, accepting this rationale would place the 'règne animal' before the 'règne végétal' and Rouelle himself admits as much. He decided to present the 'règne végétal' first however 'due to its place in the chain of phenomena' and because it 'best demonstrates the spirit of chemical analysis'.[50] What these statements mean is not entirely clear and, as seems typical of Rouelle, there is no explanation forthcoming. Nevertheless the fact that he chose to begin his course with the substances that are more difficult to explain but easier to decompose, and then move on to the 'simpler substances', represents a very empirical approach.

48. *Cours de chimie*, p.52.
49. *Cours de chimie*, p.49.
50. *Cours de chimie*, p.49.

More rationally minded instructors might have chosen to invert this order, beginning with explanations of the simple substances and working up to the theoretically complex. Rouelle, however, obviously felt that the general structure of the course should be directed by the needs and nature of things as they are revealed by experiment. Thus the very order in which the procedures are presented is an interesting compromise between pedagogical necessity and methodological preferences.

The most important observation with regard to Rouelle's presentation of the three kingdoms of matter is that they are not divided by sharp boundaries. This indeed is one of the primary themes of his course. While certain substances are 'proper' to one kingdom rather than another, meaning that they occur most frequently within that kingdom, many (if not most) substances are to be found in two if not all three of the kingdoms, and interact freely with substances in each. Rouelle stresses this point even before he takes up any of the procedures. At the beginning of the 'règne végétal' section, he lays the foundations of what he is going to do throughout the course. First, he will decompose various vegetable substances, then proceed to combine the products with those from different decompositions to demonstrate that the analysis–synthesis method of chemistry can create entirely new combinations. Next he will decompose various animal substances and recombine these not only with other animal substances, but also with many of the products derived from the analysis of vegetables. Finally, he will break down minerals and proceed to mix the parts obtained with the products of both the 'règne végétal' and the 'règne animal'. He states that 'this crossing of the kingdoms will hopefully shed some light on the nature of these substances, and allow us to know something about mutual rapports'.[51]

The manuscript contains numerous examples of these 'mutual rapports' and of the fact that the divisions between kingdoms of matter are easily transgressed. Phlogiston (the 'inflammable principle') is found in all of the 'règnes' and it is exactly the same in each.[52] Soap has its origins in oil from the 'règne animal', 'alkali fixe' from the 'règne végétal' and 'sel marin' from the 'règne minéral'.[53] The substance known as 'Prussian Blue' is used to demonstrate that plants contain iron.[54] Many of the aggregates derived from the analysis of vegetable matter, such as certain types of oil, are also derived from animal matter.[55]

The list of such examples is virtually endless, but perhaps the best demonstration of the relationship between the kingdoms of matter is found in the section on insects:

51. *Cours de chimie*, p.49.
52. *Cours de chimie*, p.525 and 559.
53. *Cours de chimie*, p.279.
54. *Cours de chimie*, p.603.
55. *Cours de chimie*, p.276-78.

Tous les insects généralment donnent de l'acide: en cela ils différent des autres animaux et puivent être regardés comme le passage du règne animal au règne minéral qui abonde en acide; comme les plantes crucifères qui donnent beaucoup d'alkali volatil ou les plantes graminées qui contennient une très grande quantité des corps muqueux peuvent être regardées comme la passage du règne végétal au règne minéral.[56]

Thus, for Rouelle, all three kingdoms are intimately linked. Moreover they are not simply united in accordance with some linear 'chain of being' concept, but have a veritable three-way relationship. Substances within the 'règne animal' can be directly related to and combine with those found in the 'règne minéral'. The 'règne végétal' is not required as an intermediary. Matter is matter, regardless of where it is found. The three kingdoms are merely variations of one another, not discrete or absolutely separated categories.

While Rouelle puts great emphasis on the relationship between the three kingdoms of matter, and especially on chemistry's capacity to demonstrate that they are all linked, the course does advance many other important ideas. Unfortunately, interesting as many of these ideas are, it would be both impractical and unnecessary to deal with all of them here. There are, however, certain concepts and themes that must be noted, if only because they are relevant to Diderot's eventual application of chemistry.

One of these crucial themes pertains to the nature of matter. Throughout the course Rouelle attacks, by implication, the Cartesian view that the physical world can be explained simply with reference to inert particles in motion. He very rarely uses the word 'force', but it does seem apparent that his understanding of chemistry necessitates a broadening of the definition of matter. This general view is revealed in his discussion of 'menstrues'. We have seen that although he refuses to be drawn as to the cause of chemical attraction, he does state that a cause must exist and that this cause must be inherent to the molecules themselves. Furthermore, by summarily rejecting Newtonian attraction and Boyle's system of 'spikes and pores' as causal explanations for chemical affinity, Rouelle opens himself up to the notion that molecules determine their mutual preferences on the basis of some internal force or quality. 'Homogeneity of surfaces' may not explain much, but it suggests that some non-mechanical (as in non-contact action), non-external, non-gravitational force is at work.

The nature of this non-mechanical, non-external, non-gravitational quality is difficult to seize upon. As was suggested earlier, Rouelle is probably referring to the Stahlian concept of 'latus', but this does not help much, because Rouelle never explicitly defines the term. The

56. *Cours de chimie*, p.280.

clearest example in Rouelle's course of what 'latus' is and how it works is Procedure 21 of the Third Part ('règne minéral'), which deals with the combination of sulphur and a fixed alkali. According to Rouelle, the two substances 'are united by the latus of phlogiston'. However, 'as this latus is very small in the fixed alkali, the union achieved is very weak'.[57] Thus, 'latus' is invoked not only as a principle by which things are drawn to one another, it plays a part in binding them and in determining the strength of that bond.

Less vague is Rouelle's second important belief regarding the nature of matter: that molecules are absolutely immutable and indestructible. To demonstrate this he argues that a molecule of air might be separated from a particular mixture, enter into another combination and, as part of that new combination, participate in exhibiting a set of properties that it did not possess in its previous state. This molecule, however, is not destroyed, and it remains identical to itself in the sense that it can just as easily be reunited with its original partner and recover all of the previous properties.[58]

This raises another important issue, namely that actions which take place on a quantitative level can have qualitative effects. As was demonstrated on a number of occasions throughout Rouelle's course, the properties of a single molecule are not identical to the properties of the mass. When two bodies combine, the result takes on an entirely new set of characteristics, with no necessary relation to the properties that its constituents possessed as distinct entities. In other words, the whole is greater than the sum of the parts. Moreover, just as it is impossible to predict theoretically which substances will have a chemical affinity for one another, so it is not possible to anticipate what properties might be exhibited by a substance resulting from the combination of two or more bodies.

A vivid demonstration of this principle is Rouelle's argument that some organic substances contain acid. Chemists had long believed that the inclusion of acid in vegetable and animal matter was impossible, because it would inevitably harm the organism. According to Rouelle, however, this view is unfounded and indeed wrong. He maintains that acid is a constituent of certain organic substances, such as mucous bodies, but since it exists there in a state of combination, it has no ill effects. The molecules of acid, while not destroyed, are no longer acting like molecules of acid.[59]

57. *Cours de chimie*, p.375.

58. *Cours de chimie*, p.25.

59. According to the manuscript, mucous bodies, such as honey or arabic gum, are of the order of 'composés'. They result from the combination of an acid (nitric acid being the type proper to the 'règne végétal') and an oil, united to a 'very subtle earth', a lot of water, and air. See *Cours de chimie*, p.124 and 242-43.

An important corollary of the idea that states of combination affect the character of substances is that two bodies which normally have no affinity for one another can be constituent elements of the same substance through the use of an intermediary. For example, substance A might have no natural rapport for substance B, but if both substances are compatible with substance C, then the three together may possibly be able to combine to form an entirely new substance, D. Of course, substance D could behave in a fashion totally different from A, B, C or even AC and BC. Making use of intermediaries in this way is fundamental to chemistry; many of the procedures performed by Rouelle in his course would not otherwise work.

Despite the overwhelmingly qualitative nature of Rouelleian chemistry, the course is not bereft of quantitative aspects. Achieving the proper ratio of combination is extremely important. Rouelle warns his students several times that neglecting proportionality can lead to failed experiments or, even worse, extremely dangerous situations. Measurement of products, either by weight or by volume, is also of interest. An example of this is Rouelle's aforementioned work on refining Hales' machine for determining the amount of air given off by a decomposing body.

On a more theoretical level, Rouelle brings quantitative considerations to bear on his famous work involving salts. Each type of salt has its own precise saturation point, or level at which it can absorb no more acid.[60] These levels can be specifically determined and thus become a defining aspect of each species of salt. While Rouelle's lessons do not treat the subject of saturation points in the same level of detail as his memoirs to the Académie des sciences, they do clearly demonstrate the point that nature does not work haphazardly.

This tendency to quantify, slight as it was, undoubtedly had something to do with the influence of physics. Indeed a number of scholars have posited that the rise of quantification constitutes the primary impact of physics on eighteenth-century chemistry.[61] A brief perusal of the manuscript however indicates that there was much more to the relationship than this. Heat, light, density, electricity, specific versus absolute gravity and even optics are all dealt with in Rouelle's lectures. This demonstrates, at the very least, that even as early as the 1750s chemists were in the position of having to assimilate and formulate opinions regarding issues that traditionally fell within the realm of physics.

What is most interesting to observe, however, is that when Rouelle discusses physics he often adopts what is best described as a 'political' tone. He is keen to point out exactly where physicists have gone wrong in their interpretation of certain phenomena and frequently offers correc-

60. *Cours de chimie*, p.210-12.

61. Tore Frängsmyr *et al.*, *The quantifying spirit in the 18th century* (Berkeley, CA, 1990), p.265.

tions or alterations to their theories. Examples of this are manifold, occurring with regard to issues such as the dissemination of odours,[62] the ordering of fossils in the Earth's crust,[63] and the products released from the purification of mercury.[64] Thus Rouelle seems anxious to assert the independence of chemistry, to confirm its status as a science which might stand alongside physics, rather than in a subordinate position. He does this even though it becomes clear from his lectures that chemistry and physics are inexorably intertwined, sharing much in the way of both content and methodology.

Rouelleian chemistry bears a similar relationship to the realm of physiology. In detailing the characteristics of the 'règne végétal' and the 'règne animal', Rouelle makes frequent reference to Boerhaave and enters into fairly detailed discussion of various physiological processes. An example of this is his statement that vegetable matter is the mediate or immediate source of everything that lives in nature. Substances proper to the 'règne végétal' are broken down through the process of digestion and then recomposed to form animal tissue.[65]

Chemistry also overlaps with physiology in Rouelle's treatment of the key difference between organic and inorganic matter. He argues that non-organic aggregates have no particular organisation, or more precisely that that while they may exhibit regularity in terms of their figure, this is not essential to their nature.[66] One can change their arrangement without changing the nature of the body. Thus, given the correct ratio of combination and the proper conditions, 'acide marin' and 'alkali volatil' can be veritably thrown together to produce 'sel ammoniac'.[67] By contrast Rouelle maintains that organisation, rather than simply composition, is absolutely seminal to organic aggregates; it plays a key role in making them what they are. Although his discussion of the issue is not elaborate, he also makes passing reference to the notion that vegetable and animal life is animated by some vivifying principle.[68]

The course concludes with a short passage on alchemy. Outlining the history of alchemy as well as its methods and goals, this concluding chapter makes it clear that Rouelle himself was a believer in the possibility of turning base metals into gold or silver. In one respect, this admission may be seen to emphasise the anachronistic nature of Rouelle's chemistry. Undoubtedly it is part of the reason that many commentators have dismissed him and his work as irretrievably 'old fashioned' or 'pre-

62. *Cours de chimie*, p.56.
63. *Cours de chimie*, p.304.
64. *Cours de chimie*, p.462.
65. The introduction to the second part of the manuscript ('Du règne animal') includes lengthy discussion of the process of nutrition from a chemical point of view (*Cours de chimie*, p.241-42).
66. *Cours de chimie*, p.9.
67. *Cours de chimie*, p.443.
68. *Cours de chimie*, p.290.

scientific'.[69] Nevertheless Rouelle's hopes for the prospect of transmutation must be put into perspective. The most important conclusion that one can draw from it is that for Rouelle, chemistry still represented a discipline in which almost anything was possible. The bodies of nature as revealed to the senses were not regarded as the simplest substances possible. Therefore the core method of chemistry held out the possibility of being able to take any substance, break it down into its constitutive parts and then recombine these with others to create an entirely different substance. In the imagination at least, any substance could ultimately be transformed into any other.

Rouelle's course, then, covered an extremely wide variety of ideas, presumptions and phenomena. Jean Mayer claims that from Rouelle, Diderot learned everything there was to know about chemistry at that time.[70] This claim seems rather great and it may be more appropriate to posit simply that Diderot received a thorough grounding in chemistry as it was practised and understood in France during the mid eighteenth century. In any case, by attending the lecture series for three years and by composing the textbook, Diderot acquired an intimate understanding of Rouelle's chemistry. More importantly Rouelle's lessons made a significant impression upon Diderot, and excited in him an interest in chemistry that was not present prior to 1754. This interest manifested itself in a number of ways.

One testament to Diderot's new-found enthusiasm for chemistry is the nature of the activities in which he participated. At least two scholars have argued that, after having made the acquaintance of Rouelle, Diderot frequently visited the chemist's laboratory. The purpose of these visits, according to Paitre, was 'to find answers to questions raised in *Pensées sur l'interprétation de la nature*'.[71] Presumably attendance of Rouelle's course might have accomplished this, but supplementary visits to the laboratory would not have hurt. Unfortunately the terms and nature of these visits is unknown. Indeed there is no direct evidence confirming that Diderot actually visited Rouelle's laboratory at all. The entire issue, then, must remain one of speculation.

69. Of course, proposing that alchemy is somehow antithetical to the emergence of modern science is a relatively unhelpful way of approaching the complex of interests that captivated seventeenth- and eighteenth-century natural philosophers. Generally speaking, in the seventeenth and eighteenth centuries belief in alchemy was not perceived to be at odds with good natural philosophy. Perhaps the best testament to this is the fact that many of the greatest figures in the development of early modern science maintained an avid interest in the discipline, see for instance Betty Jo Dobbs *The Janus face of genius: the role of alchemy in Newton's thought* (Cambridge, 1991), John Henry, *Knowledge is power; how magic, the government and an apocalyptic vision inspired Francis Bacon to create modern science* (Cambridge, 2002) and Lawrence Principe, *The Aspiring adept: Robert Boyle and his alchemical quest* (Princeton, NJ, 1998).

70. See Mayer's critical edition of *Eléments de physiologie*, p.lvi.

71. Fernand Paitre, *OC* (Lyon, 1904), p.22.

More is known however regarding the ongoing association of Diderot with Rouelle's younger brother (H. M. Rouelle, known to his contemporaries as Rouelle 'le cadet') and the great chemist's son-in-law Jean d'Arcet (1724-1801). Even after the elder Rouelle's death in 1770, Diderot maintained a fruitful association with both of these men. During his trip to Russia, for instance, Diderot was entrusted with collecting samples of rare minerals in order that d'Arcet might carry out experiments on their chemical characteristics.[72] More significantly Diderot's *Correspondance* reveals that he was able to gain access to d'Arcet's laboratory on several occasions. The most famous of these was in 1771 when Diderot was invited, along with a host of Europe's most notable scientists, to observe an experiment on the volatility of diamonds. Diderot recounted this experiment in a piece called *Expériences intéressantes*,[73] which was published in Grimm's *Correspondance littéraire*. Such experiences allowed Diderot to witness at first hand the day-to-day work of experimental science and provided him with the invaluable opportunity to test and to verify his own ideas about scientific method. Jean Mayer once proclaimed that Diderot studied scientists as they in turn study nature; the *philosophe*'s experience of observing d'Arcet at work would have certainly facilitated this project.

While these experiences are undoubtedly significant, the best evidence that Rouelle's course impelled Diderot to take an interest in chemistry is revealed by an analysis of his writings. The subject of chemistry is almost completely neglected in Diderot's pre-1754 work, but ideas related to the discipline appear very frequently in his later thought. One obvious example of this is Diderot's *Lettre d'un citoyen zélé*. The *Lettre* is an editorial on the debate that raged in France for most of the eighteenth century between the medical and surgical professions.[74] Nowhere in the original version (1748) of the work is there any mention of chemistry. In 1760 however (three years after completion of Rouelle's course), Diderot added two paragraphs, both of which deal exclusively with chemistry. Extrapolating upon Rouelle's proclamation that chemistry is indispensable to medicine, Diderot asserted that both doctors and surgeons should cultivate a greater knowledge of pharmacy.[75]

The *Encyclopédie* also bears witness to the rising importance of chemistry

72. Diderot, *Correspondance*, vol.16, (June 15, 1774), p.44.

73. Diderot, *OC*, vol.9, p.880-89.

74. This controversy subsumed a wide variety of issues, including the corporate and academic status of surgeons, the value of their discipline relative to that of the doctors, the respective rights of either group to indulge in given practices, and, certainly not least, prestige. For a summary of this great (and, for a time, very public) debate, see Laurence Brockliss and Colin Jones, *The Medical world of early modern France* (Oxford, 1997), p.590-605. For more on Diderot's appraisal of the situation, see chapter 5 p.188-89.

75. *Lettre d'un citoyen zélé*, *OC*, vol.2, p.148.

in Diderot's thought after the mid 1750s. In the *Prospectus* (June 1751) chemistry is almost completely neglected. While fairly substantial sections are devoted to the nature and place of other scientific disciplines within the 'tree of knowledge', chemistry receives only one (relatively short) paragraph. This brief, almost token paragraph simply points out the practical value of studying chemistry, largely on account of its mineralogical applications.[76]

By contrast, Diderot's 1754 article 'Encyclopédie' (written during the first year that he attended Rouelle's lectures) devotes much more space to the subject of chemistry and presents it in a much more positive fashion. Chemistry is mentioned on no less than six separate occasions in the article and, in certain contexts, receives more attention than any of the other sciences. More importantly, in the article 'Encyclopédie' Diderot demonstrates a much greater awareness of what chemistry is and of what basic operations are at its heart.[77] Chemistry is acknowledged not simply as an art, but as a science that is crucial to unlocking the secrets of nature.

The place that chemistry was to assume in Diderot's thought after his exposure to Rouelle is perhaps best demonstrated by *Plan d'une université*. This work refers to chemistry as 'la rivale du grand ouvrier'.[78] It argues that chemistry, along with mechanics and natural history, is one of the most important disciplines to understand. Diderot portrays chemistry as the primary means of perfecting physics and of enlightening the mechanical arts. The section on medicine argues that chemistry is indispensable to the profession and that it is the primary means of elaborating upon the three kingdoms of matter laid out by natural history.[79] Also of note is the fact that the reading Diderot recommends as far as chemistry is concerned is nothing other than a textbook of Rouelle's course.[80]

Noting the references in *Plan d'une université*, the article 'Encyclopédie' and *Lettre d'un citoyen zélé* is merely to scratch the surface how Diderot's post-1754 canon reflects the increased prominence of chemistry in his thought after attending Rouelle's course. Almost every Diderot text that was written after his exposure to Rouelle and which has anything to do with natural philosophy betrays the influence of chemistry, or at least makes some mention of the discipline. This applies to *Le Rêve de D'Alembert* (1769), *Principes philosophiques sur la matière et le mouvement* (1770), *Réfutation d'Hemsterhuis* (1774), *Voyage en Hollande et dans les Pays-Bas autrichiens* (1775), *Réfutation d'Helvétius* (disseminated in the *Correspondance littéraire* from 1783 to 1786) and *Eléments de physiologie* (1778-1780).

76. *Prospectus*, p.310.
77. 'Encyclopédie', p.365-463.
78. *Plan d'une université*, p.781.
79. *Plan d'une université*, p.825.
80. *Plan d'une université*, p.782.

It is one thing, of course, to identify the trend, quite another to interpret it. While the textbook of Rouelle's course provides a fairly neat picture of the information that was presented to Diderot in the 1750s, it is an entirely different and more difficult task to understand exactly what Diderot did with this information. In attempting to unravel the role of chemistry in the development of Diderot's thought, we might first go back to *L'Interprétation de la nature*. This may seem a curious thing to do, given that *L'Interprétation de la nature* was written in 1753, but there are a number of interesting parallels between the scientific method that Diderot advocates in that work and Rouelle's manner of investigating nature. First, Diderot's ideas about the primacy of experiment and the need for a careful, patient, empirically based approach to interpreting nature are mirrored in Rouelle's lessons. The introduction to the course contains a passage very reminiscent of *L'Interprétation de la nature* which states that the chemist needs finesse and good instruments, and that the progress of chemistry is so slow because 'either we don't have the instruments or we ignore their just application'. It goes on to state that 'our impatience keeps us from focusing our attention on the same object for a long time – which is what we need to do'.[81] Of course, if these lines have a very Diderotesque ring, it might well be due to the fact that he penned them. Nevertheless, Rouelle did make the point.

A second, less explicit example of concurrence between the scientific methods of Diderot and Rouelle are the former's ideas about the ideal scientist letting himself be guided by Nature, and about the intuitive 'man of genius'. Although Rouelle may not have regarded himself as a 'man of genius', there is evidence that Diderot did. This is apparent in a 1770 obituary of the chemist in which Diderot lauds Rouelle for possessing many of the qualities that are enthroned in *L'Interprétation de la nature*. Referring to Rouelle as a 'grand savant', Diderot says that 'il réunissait le don du génie à une érudition profonde'.[82] He also says of Rouelle that 'Il appliquait ses expériences au système général du monde; il embrassait les phénomènes de la nature et les travaux des arts; il les liait par les analogies plus fines; il se perdait, on se perdait avec lui, et l'on ne revenait jamais à l'objet particulier de la démonstration du jour, sans être étonné de l'espace immense qu'on avait parcouru.'[83] It seems that the manner in which Rouelle taught was very similar to the manner in which Diderot wrote *L'Interprétation de la nature*, simply letting the thoughts flow from his pen as they presented themselves to his mind. This kind of intellectual flexibility, united to an empirical approach, is at the very centre of Diderot's notion of the scientific genius. Although he downgraded the role in natural philosophy of abstract reasoning and

81. *Cours de chimie*, p.10.
82. *Note historique*, p.602.
83. *Note historique*, p.600.

unsubstantiated speculation, Diderot held dear the notion of playing with hypotheses and of regarding the exploration of nature as something of an adventure. His description of Rouelle seems to portray a man of very much the same mind.[84]

Thus, rather than supplying Diderot with an idea of proper scientific method, Rouelle merely confirmed that which already existed for the *philosophe*. If the experience of attending Rouelle's lectures contributed at all to Diderot's views on methodology, it was only in terms of vindication and the provision of further compelling evidence that he had been right all along. Of course, we should not underestimate how important and reassuring it might have been for Diderot to have his own suspicions and prescriptions for proper experimental method confirmed and demonstrated by one of the most highly esteemed practising scientists of the day. In any case, given the immediate concurrence between Diderot and Rouelle on matters of how scientists should conduct themselves, it is unsurprising that the former was so well disposed to accept many other ideas of the latter, and that chemistry in general should have made such an impact on Diderot's thought.

The specific nature and terms of this impact centred around conceptions of matter and its mode of operation. Although Diderot had devoted considerable thought to such issues prior to attending Rouelle's lectures – the need to broaden the accepted definition of matter had been one of the primary themes of *L'Interprétation de la nature* – his exposure to the chemistry of Rouelle assisted Diderot enormously in this project by lending new impetus, justification and terminology. Where Diderot had earlier addressed the need for natural philosophers to alter their perception of matter in vague and metaphysical terms, he would later (post-Rouelle) speak in terms that might be called physico-chemical. *L'Interprétation de la nature* is predominantly negative in tone; it is largely concerned with rejecting the ideas of others (such as those of Maupertuis) as they pertain to the nature and behaviour of matter. Where the work does seek to advance notions about how matter may actually be constituted it manifests, as we have seen in chapter 2, the influence of metaphysical, Leibnizian ideas such as *vis viva*, or 'living force'.[85] By contrast, works written after 1754, such as *Matière et mouvement* (written no earlier than 1770, although not published until 1792), are thoroughly imbued with terms, concepts and examples drawn from Rouelle.[86] Not

84. This has been noted by John Hope Mason. See Mason, *The Irresistible Diderot*, p.267.

85. *Pensées sur l'interprétation de la nature*, p.755-60.

86. As was pointed out at the beginning of this chapter, Arthur Wilson has observed that *Matière et mouvement* 'goes distinctly beyond his earlier speculations regarding matter in *Pensées sur l'interprétation de la nature*. The difference could have been brought about by the fact that meanwhile Diderot had attended three years of Rouelle's lectures in chemistry' (Wilson, *Diderot*, p.580). The pages that follow will demonstrate why Wilson's suspicion is indeed correct.

only did the operations carried out before Diderot's eyes at the Jardin du Roi impress even more firmly upon him the fact that the old paradigm of a universe explicable solely in terms of inert matter and motion was untenable, but knowledge acquired from Rouelle enabled Diderot to frame his quest to broaden the definition of matter in a more forceful, integrated and scientific manner. Rouelle lent Diderot the concepts necessary to make a more convincing argument, and helped to give very definite shape to the way in which Diderot would regard matter.

In elucidating this enhanced concept of matter it makes sense to begin with the basic nature of molecules themselves. Rouelle taught, as we have seen, that the molecules which combine to make up the substances of the world possess absolute integrity. Molecules cannot be divided or destroyed, and although the properties they exhibit upon entering into certain states of combination may change, they themselves remain true – if extracted from the state of combination they revert to exhibiting exactly the same qualities that they did before becoming part of some larger body. Diderot picked up directly upon this lesson, asserting that molecules are 'immuables, éternelles, indestructibles'.[87] He conceived of molecules as being the basic units of matter, and further that these units were fundamentally unalterable, regardless of the travails to which they may be subjected by either art or nature.[88]

Although opposed to the presumption held by many Cartesians that matter was indefinitely divisible, these basic assertions were not revolutionary. Where Diderot's view of matter emerged as genuinely unorthodox was in its insistence that motion was inherent to matter itself. In *Matière et mouvement* Diderot argued that there were three types of force which governed the behaviour of matter: the force of gravity, external forces (the action of all other molecules on the one in question) and the internal force of the molecule itself.[89] This last type of force Diderot imagined as being essential to the molecule, as crucial to its identity as extension or impenetrability. To further drive an intellectual wedge between external forces acting on the molecule from without and internal forces intimate to it, Diderot remarked that 'La force qui agit sur la molécule, s'épuise. La force intime de la molécule ne s'épuise point. Elle est immuable, éternelle. Ces deux forces peuvent produire deux sortes de *nisus*; la première un *nisus* qui cesse; la seconde un *nisus* qui ne cesse jamais. Donc il est absurde de dire que la matière a une opposition réelle

87. *Matière et mouvement*, p.614.

88. It should be noted that Diderot employs the term 'molecule' quite loosely, and often uses it interchangably with the term 'atom'. That his notion of both or either of these words is far removed from twenty-first-century usage goes without saying. What is important here is simply that when Diderot does refer to 'molecules' or 'atoms' he is referring to the most basic units of matter imaginable.

89. *Matière et mouvement*, p.613.

au movement.'[90] Therefore even the contributions of Newton, which Diderot acknowledged like so many of his contemporaries to have shed an unprecedented amount of light on the observed phenomena of the universe, were regarded by the *philosophe* as merely scratching the surface of questions pertaining to the nature and operation of matter. In *Réfutation d'Hemsterhuis*, Diderot remarked that 'l'attraction n'est pas la seule force, ni même peut-être la principale des forces qui agissent dans la nature'.[91] This statement paralleled his declaration in *Matière et mouvement* that the force of gravity ('cette force générale et commune') is contradicted by one hundred thousand others.[92]

Amongst these 'one hundred thousand other forces' was 'chemical force' or 'chemical vitality', which manifested itself in nothing other than chemical affinity, or the tendency for a substance to combine with one rather than another of its fellow types of matter. Some of Diderot's contemporaries believed that this affinity was equivalent to, or at least explicable in terms of attraction after, the Newtonian model. Buffon fell into this category, and more notably so did a number chemists that studied under Rouelle, including Macquer, A. Baumé (1728-1804) and Guyton de Morveau (1737-1816). These three did much to promote the idea that Newtonian attraction might serve as the foundation of affinity theory, although the idea did have a relatively long and noble heritage, much like the proposition that the preferences for molecular combinations were governed by mechanical considerations such as the shape or texture of the molecule's surface. These systems were, as we have seen, both referred to and explicitly rejected by Rouelle in his lectures on chemistry. Diderot too spurned such oversimplifications. In *Réfutation d'Hemsterhuis*, immediately subsequent to the aforementioned assertion that gravity may not even be the principal force at work in nature, Diderot stated that 'Il y a la combinasion, et j'entends par ce mot le principe d'affinité des chimistes, qui est tout autre chose que l'attraction'.[93] In other words, Diderot was addressing an entirely new species of force, something to be ranged alongside gravity and mechanically imparted action as crucial to understanding the behaviour of matter.

It should be noted at this point that Rouelle himself did not explicitly state that movement is inherent to matter. As Rhoda Rappaport has observed, Rouelle did not attempt to explain the attractive and repellent forces which governed chemical affinity but, in the Newtonian manner, accepted the observed phenomenon of (chemical, as well as gravitational) attraction as an unknown but essential property of matter.[94] It would

90. *Matière et mouvement*, p.611-12.
91. *Réfutation d'Hemsterhuis*, p.49.
92. *Matière et mouvement*, p.612.
93. *Réfutation d'Hemsterhuis*, p.49.
94. Rappaport, 'Rouelle and Stahl', p.75.

seem then that Diderot went, or wanted to go, farther than Rouelle in asserting the reality of internal forces in general and chemical force in particular. This does not pose a serious challenge to the suggestion that Diderot was influenced by Rouelle in elaborating a theory of matter however, because Diderot himself did not attempt to explain the true nature of these forces any more than Rouelle. Just as Newton regarded with suspicion any attempt to offer an explanation for the true cause of gravity, so Diderot was averse to the idea of trying to account for the true cause of chemical affinity. In both cases, it was enough to observe these respective forces at work and to describe their effects; explanations of first causes were not required to believe that such forces actually existed. Therefore, Diderot's belief that molecules possess their own, innate forces, and that these forces are responsible for the phenomenon known as chemical affinity, is not actually that distinct from Rouelle's methodologically circumspect treatment of the same issue.

The only other question mark surrounding the issue of Rouelle's influence on Diderot's notion that force is inherent to matter is the assertion launched by Lester Crocker that Diderot actually inherited the idea from the English philosopher John Toland (1670-1722).[95] Toland, it is true, did advance the idea that matter is intimately endowed with energy, and Diderot did read *Letters to Serena*, the work in which Toland proclaimed this.[96] The problem with Crocker's interpretation however is that it is not known exactly when Diderot read Toland's piece, whereas we obviously know exactly when and how he was exposed to the thought of Rouelle. Furthermore the ideas of Diderot and Toland differed in a very important respect: Toland, despite his conviction that motion was inherent to matter, felt that all matter was nevertheless homogeneous. This assumption could not have been further removed from Diderot's own perception. That matter was heterogeneous, rather than homogeneous, was one of the positions that Diderot defended in a most lively manner, and it is arguably the primary theme of *Matière et mouvement*.

Put simply, Diderot proclaimed that the immutable, eternal, indestructible molecules which comprise matter are not all alike but differ in terms of kind and quality. To posit otherwise, according to Diderot, was an absurdity which amounted to abstracting molecules out of existence.[97] Amidst the infinite number of molecules in the universe, Diderot asserted that we find an infinite number of different natures. Each particle is of a type, be it fiery, aqueous, nitrous, alkaline or sulphurous.[98] These characteristics, of course, could be detected through chemical analysis, by

95. Lester Crocker, 'J. Toland et le materialisme de Diderot', in *Revue d'Histoire littéraire de la France* 53 (1953), p.289-95.

96. John Toland, *Letters to Serena* (London, printed for Bernard Lintot, 1704).

97. *Matière et mouvement*, p.611.

98. *Matière et mouvement*, p.611.

breaking substances down as far as they will go, by observing the properties of each different type of matter, and by playing with various combinations of matter to see what kind of qualities they exhibit. That Rouelle's chemistry gave shape and substance to Diderot's assertion that matter is heterogeneous is amply demonstrated in the following passage from *Matière et mouvement*, which is sweeping, energetic and uncompromisingly expressed:

> Mais j'arrête mes yeux sur l'amas général des corps; je vois tout en action et en réaction; tout se détruisant sous une forme, tout se recomposant sous une autre, des sublimations, des dissolutions, des combinaisons de toutes les espèces, phénomènes incompatibles avec l'homogénéité de la matière: d'où je conclus qu'elle est hétérogène; qu'il existe une infinité d'éléments divers dans la nature; que chacun de ses éléments par sa diversité a sa force particulière, innée, immuable, éternelle, indestructible; et que ces forces intimes au corps ont leurs actions hors du corps; d'où naît le mouvement ou plutôt la fermentation générale dans l'univers.[99]

This vision recalls Rouelle's lesson about the individuality of characteristics exhibited by one type of molecule as opposed to another, and in setting forth the argument it also makes direct reference to aspects of Rouelle's chemistry such as decomposition and recomposition, dissolutions, and the immutable, eternal, indestructible nature of molecules themselves. The passage even uses fermentation, a chemical process described by Rouelle in great depth, as a metaphor for the operation of the universe as a whole.

The thrust of this argument ties in well with Diderot's conviction in the need to respect the particularity of things. If matter is seen as heterogeneous, and the behaviour of molecules, which amounts to the most basic movement of the universe, is governed not simply by quantitative, measurable, external considerations but by qualities, properties or forces inherent to the molecules themselves, then greater emphasis is put on descriptive science. The scientist cannot rely on generalisations to arrive at the truth, but must take the time to carefully observe each aspect of reality. Just as Rouelle asserted that the chemical affinity or 'rapport' of substances for one another can only be discovered through trial and error, so Diderot argues 'Il y a autant de lois diverses qu'il y a de variétés dans la force propre et intime de chaque molécule élémentaire et constitutive des corps'.[100] The universe is regarded as an organism that can only be comprehended by taking it apart piece by piece, by dissolving it into its constituents. For both Rouelle and Diderot, the theories surrounding what matter is both legitimise and necessitate a very particular method for the means by which it is to be studied.

This emphasis on heterogeneity and particularity has been seen by

99. *Matière et mouvement*, p.613-14.
100. *Matière et mouvement*, p.612.

some, such as Jean Mayer, to go hand in hand with an anti-mechanistic approach to natural philosophy. But the anti-mechanistic impression that sometimes comes across in Diderot's works was not necessarily meant to imply that the universe is lawless or chaotic. Indeed, anti-mechanism may be the wrong adjective to apply to Diderot's philosophy; supra-mechanistic would be more accurate. The intention of Diderot's chemistry was not to deny the laws of mechanics, but merely to suggest that there are more subtle, complex laws of reality which operate on a much finer scale than those to which the dual principles of (inert) matter and motion can be applied. This is an important point to stress, for it affirms that for Diderot there were laws of chemistry. Certain combinations of substances would invariably lead to the creation of particular products or events. That Diderot acknowledged this is revealed in *Matière et mouvement*, which states that one 'ne se trouve point à côté de trois molécules combinées de salpêtre, de charbon et de soufre, sans qu'il s'ensuive une explosion *nécessaire*' (*italics* are mine).[101] The notion that arises from Diderot's work is that the laws of chemistry consist in nothing other than the various 'rapports' or relationships that actually exist between different types of substances, but that these relationships are nevertheless regular and governed by the unalterable nature of the molecules themselves. Thus in *Matière et mouvement*, a work in which Diderot refers to himself as both 'physicien et chimiste',[102] he upholds what might be called both the Newtonian and the Rouellian paradigms regarding the behaviour of matter. In addressing the action of homogeneous aggregates upon one another Diderot maintained that the force of their actions and reactions are in direct proportion to their mass.[103] In doing this, he was reinforcing the regular, mathematical, Newtonian world-view. In terms of heterogeneous aggregates however he argued, as has been witnessed, that the end result is not necessarily the same (as the one that would be predicted by Newtonian mechanics) because there could be (and often are) other laws and forces at work.[104] Thus it was that Diderot could posit that matter is governed by three distinct types of force (gravitational, externally imparted and internal) and moreover that it was not the laws of movement which required revision, but simply the conception of matter.[105] The laws of the universe may have been, according to the Diderotian conception, very diverse, but they existed nonetheless.[106]

101. *Matière et mouvement*, p.611.
102. *Matière et mouvement*, p.611.
103. *Matière et mouvement*, p.612.
104. *Matière et mouvement*, p.612.
105. *Matière et mouvement*, p.610.
106. There is a question here as to how literally Diderot should be taken when he refers to 'infinite' substances and laws etc. As I. Bernard Cohen observed, the proposition that for each aspect of the universe there applies a different law is to posit no explanation of things whatsoever. See Cohen, *The Newtonian revolution*, p.100. The primary point here is Diderot's

It is still worth remarking, however, that the diversity of phenomena, actions and molecules that Diderot justified on the basis of Rouelle's chemistry enabled him to posit a certain unpredictability at the heart of nature or, better, to cede to nature a significant element of mystery. There is the sense in reading Diderot that he occasionally cherishes the thought of a universe that, though essentially 'noble' in that it must obey rules, is not only stranger than we imagine, but stranger than we are capable of imagining, and that is perpetually able to present its investigator with intriguing surprises. This kind of sentiment is reflected in passages such as 'je ne serais point surpris qu'il y eût dans la nature, telle molécule qui, jointe à une autre, rendît le mixte résultant plus léger'.[107] The perplexed sense of wonder that Diderot exhibits here on account of nature's unpredictability may have even drawn its inspiration from Rouelle's lesson on the calcination of antimony, in which the chemist ponders the curious phenomenon of how the substance is reduced in volume but gains in weight.[108] In any case, Diderot's statement certainly reprises the critical Rouellian idea that the properties possessed by the product of a combination bear no necessary resemblance to those of its constituent ingredients. Qualitatively, the whole can be greater than the sum of its parts. To put things another way, quantitative change can and does lead to qualitative results. A corollary of this, for Diderot just as for Rouelle, was that isolated molecules have none of the properties of the mass. Diderot acknowledged this explicitly in a number of works, most notably *Eléments de physiologie*, which shall be discussed in greater depth below.

Another key lesson that Diderot received from Rouelle pertained to the kingdoms of matter and their interaction. While much of Diderot's early thought was influenced in this respect by the work of Buffon, later expositions bear the hallmark of Rouelle. Buffon placed a firm boundary between the organic and inorganic realms of matter, whereas for Rouelle the situation was much more fluid. The passage on insects being intermediaries between the 'règne animal' and the 'règne minéral', as well as frequent demonstrations of the inorganic (metallic or even acidic) substances present in organic compounds, opened the door for Diderot to be able to assert that the linkage of the kingdoms of matter was far more intimate than Buffon may have cared to guess. Rouelle broadened and deepened Diderot's understanding of the 'règnes', showing him that they participated in a three-way relationship (rather than a linear, 'bottom-to-top' relationship), that they were inexorably intertwined and that the

insistence that we are living in a multiple force universe – one that is ultimately ordered, but which is almost inconceivably more complicated than that provided by Newton, much less Descartes.

107. *Matière et mouvement*, p.614.

108. *Cours de chimie*, p.519.

elements of each category were relatively interchangeable. In other words, Rouelle introduced for Diderot a move away from the natural historical classification of the kingdoms of matter towards an understanding informed by chemistry that was more flexible, more intimate to the nature of matter itself, and which involved a much more delicate interplay.

All of this lays the groundwork for what may be the most important element of Diderot's understanding of matter: that it can change from one state to another, and that the impetus for this change can come from within. This idea is crucial to the concept of *sensibilité universelle* as outlined in *Le Rêve de D'Alembert*. Although an altogether different kind of work than *Matière et mouvement*, the *Rêve* embraces the same basic theme. It is a poetic expression of the notion that reality is constituted of matter which is eternally active, infinitely diverse and capable of becoming anything. While traditional readings of the *Rêve* emphasise its indebtedness to the biological thought of the Montpellier vitalists, and while such interpretations are no doubt valid, comparisons with *Matière et mouvement* reveal that it is also underpinned by chemistry.[109] Armed with Rouelle's concepts of the molecule, the autonomy of chemical operations and the possibility for matter to undergo wholesale changes of quality without individual atoms being irrevocably compromised, Diderot was able to portray the universe as unified and matter as transformable without resorting to animism (the proposition that matter itself is inert but is somehow moved by 'the soul', 'animal spirits' or 'occult forces' that are something separate from matter itself).[110]

The chemical underpinnings of *Réfutation d'Hemsterhuis* are even more obvious. In making the materialist argument that the 'soul' is entirely dependant upon the flesh, Diderot made frequent appeals to lessons learned from Rouelle. In addition to discussing chemical affinity and the inherence to matter of force or energy, *Réfutation d'Hemsterhuis* is littered with references to chemistry. Among these references are statements about gluten (a substance extensively studied by Rouelle), the chemical analysis of plants, the argument that living molecules are the product of ferment and putrefaction, and that the perpetual flux of the universe depends upon the heterogeneity of molecules.[111] It now becomes fully apparent that the demonstrations and concepts presented to Diderot

109. This was acknowledged by Aram Vartanian in his last work, published posthumously. Regarding the story of how a marble statue may be turned to flesh, Vartanian noted: 'Diderot's aim is to show that the "animalisation of matter" is no inconceivable mystery or miraculous creation, but a familiar sequence of chemical reactions which it is possible, not only to observe and describe, but also to initiate and control'. See Vartianian, *Science and humanism in the French Enlightenment*, p.169. The relevant section of *Le Rêve de D'Alembert* is to be found in *OC*, vol.8, p.59.

110. For more on these issues, see chapters 4 and 5.

111. *Réfutation d'Hemsterhuis*, p.7-105.

at Rouelle's course enabled the *philosophe* to propound as plausible, empirically rooted scientific hypotheses such ideas as the almost unlimited transformability of matter and the material basis of the soul.

The relevance of Rouelle's teachings to the thought of Diderot does not end here however. There are at least two other respects in which Rouelle performed for Diderot the role of mentor. The first of these involved the place of measurement in experimental philosophy. While Rouelle never fully exploited, indeed never realised the profound significance of precise measurement in the way that Lavoisier would do, he did make use of the chemical balance and hastened to point out, as we have seen, that measurement was to be employed in the conduct of experiments.[112] Something of a rule of proportionality seems to have been vaguely nascent in his thought, and it would be the same with Diderot. That this was the case is revealed in the aforementioned piece *Expériences intéressantes*. In relating the experiment and the manner in which it was conducted, Diderot assiduously recorded details of weight and time. He did not, however, make tremendous use of his recordings, and while these details do help to frame the experiment, they are not truly interpreted.[113] In other words, Diderot pushed his concerns regarding careful measurement to about the same degree as Rouelle – no more, no less. Of course, it should be granted here that the manner in which Diderot acknowledged the role of weighing substances in the course of an experiment, and the fact of his paying close attention to the time that various minerals spent in the furnace need not have come exclusively from Rouelle. Indeed it may be presumed that the man carrying out the procedure, Jean d'Arcet, was making exactly the same recordings as Diderot.

Less circumstantial however is the final respect in which Rouelle's ideas were brought to bear upon the work of Diderot, that is, the illumination and nuancing of the latter's physiology. Jean Mayer argues that the influence of Rouelle on Diderot's *Eléments de physiologie* (composed over the course of the 1770s) is hard to trace, but there is no doubt that the work bears the stamp of chemistry and that references to the discipline are scattered throughout.[114] These references are of several different orders. Certain passages related to chemistry are simply general statements of theory that may or may not underpin arguments that Diderot is trying to formulate. In other instances, it is clear that Diderot is

112. For more on the role and development of precise measuring techniques and quantification in chemistry during the eighteenth century, see Anders Lundgren, 'The Changing role of numbers in 18th-century chemistry', in Frängsmyr *et al.*, p.245-65.

113. *Expériences intéressantes*, *OC*, vol.9, p.881-87.

114. *Eléments de physiologie*, Mayer edn, p.li. Mayer's rather pessimistic appraisal has more recently been adjusted however, by Paulo Quintili, whose recent critical edition of *Eléments de physiologie* acknowledges that Rouelle helped to furnish Diderot with ideas regarding the inherence of energy to matter and also bolstered the *philosophe*'s ideas about universal dynamism. See *Eléments de physiologie*, ed. Paulo Quintili (Paris: 2002), p.16-18.

expounding a physiological view that is directly informed by chemistry. Finally, there are references to chemistry that seem random and disconnected but which hold appeal if only because they stand as testimony to the fact that Diderot was still very preoccupied with chemistry more than twenty years after the course of Rouelle.

The first category of reference is applicable primarily to statements from the introduction to the *Eléments*. Included is a passage about how isolated molecules possess none of the properties of the mass.[115] This notion, inherited directly from Rouelle, is an important acknowledgement of the idea that an object can be more than the mere sum of its parts. Later Diderot would apply an analogous argument to the debate on generation, claiming that it is not necessary for a germ to resemble the animal that it will become.[116] Chemical operations form the paradigm for the nature of growth, change and life itself. Also falling into this class of reference are statements regarding issues such as the transformation of matter, the notion of 'vital chemical energy' and the four chemical principles proposed by Rouelle (air, earth, fire and water). These ideas do not always contribute directly to the progress of the work as a whole, but they do assist in developing the background for Diderot's physiology.

The direct impact of chemistry on the *Eléments* is more apparent elsewhere. In many instances Diderot interprets matters that are distinctly physiological from a chemist's point of view, or brings chemistry to bear on issues that might otherwise be inexplicable. Subsumed under this class of reference is Diderot's aforementioned classification of the 'règnes' in terms of chemistry. A good example of the manner in which he does this is the statement: 'Le champignon donne à l'analise de l'alkali volatil signe caractéristique du règne animal.'[117] Mayer criticises Diderot here for lacking clarity of thought, but commends the *philosophe* for being on relatively novel ground in discussing the kingdoms of matter in terms of chemical analysis. It is, however, a notion inherited from Rouelle, who taught that volatile alkalis are 'proper to the animal kingdom, just as nitric acid is proper to the vegetable, vitriolic acid is to the mineral, and marine salt is to substances found in the sea'.[118]

Heat and fermentation are two other aspects of Rouelle's course that Diderot applies to the *Eléments*. The clearest instance of this is Diderot's contention, already witnessed in part with regard to *Réfutation d'Hemsterhuis*, that heat and fermentation are crucial to the production of vegetable matter. In a related discussion he addresses the subject of how vinegars, acids and salts are all produced from fermentation, arguing that water destroys the superabundant 'nisus' (being an effort towards

115. *Eléments de physiologie*, Quintili edn, p.107.
116. *Eléments de physiologie*, Quintili edn, p.134.
117. *Eléments de physiologie*, Quintili edn, p.115.
118. *Cours de chimie*, p.260.

conservation) of certain bodies, thus isolating the particles and rendering them active.[119] Although Rouelle did not actually use the word 'nisus' in his lectures, he did speak at length about fermentation and emphasised the roles of both heat and water in exciting the process.[120]

The section of the *Eléments* from which the above examples are drawn (entitled 'Végéto-animal') also makes reference to work conducted by Rouelle and Macquer on determining the constituents of wheat. Separating the starch from the gluten, Rouelle concluded that starch is proper only to what may be called the vegetable part of wheat, but that gluten might be termed 'végéto-animal'.[121] This experiment obviously brings to mind the controversial work of John Turberville Needham (1713-1781) and indeed Diderot mentions the Rouelle/Macquer theory largely by way of introduction to a discussion of Needham's 'eels'.[122]

Perhaps Diderot's most interesting application of chemistry to matters of physiology is his description of various elements of the body in distinctly chemical terms. This approach is apparent in virtually every chapter of the book. For example, in the section on fibres Diderot proclaims that gluten, the fluid element of fibre, contains water, salt, air and oil.[123] He characterises blood by very similar means, listing its constituents as sea-salt, oil, iron, non-elastic air and electrical material (the presence of which is proven, according to Diderot, by its odour and luminescence).[124] He goes on to insist that the elements in combination form something fundamentally different in nature from the ingredients themselves. This, of course, reprises a theme found in *Matière et mouvement* and was one of Rouelle's most significant lessons.

The section on humours, secretions and glands acknowledges the physiological use of chemistry-related processes such as analysis, dissolution and evaporation.[125] It refers to the extraction of insipid carbon from the boiling of humours, and to the chemical constitution of different types of secretion (water, salt, oil and 'concrete volatiles' such as ammonium).[126] Intriguingly the chapter also notes that one of the principal means by which humours are secreted is by chemical affinity.[127] Diderot fails to elaborate upon this idea here, but it does serve as evidence that he

119. *Eléments de physiologie*, Quintili edn, p.110.

120. *Cours de chimie*, p.130-36.

121. *Eléments de physiologie*, Quintili edn, p.111-12.

122. J. T. Needham was an English microscopist who aroused tremendous controversy throughout Europe after claiming to possess empirical evidence that life can arise from non-life. For more on the nature of his 'discovery' and the manner in which it was interpreted, see chapter 4, p.138.

123. *Eléments de physiologie*, Quintili edn, p.156.

124. *Eléments de physiologie*, Quintili edn, p.193.

125. *Eléments de physiologie*, Quintili edn, p.203-10.

126. *Eléments de physiologie*, Quintili edn, p.204.

127. *Eléments de physiologie*, Quintili edn, p.203.

is employing the concepts of chemistry to solve or to understand physiological problems.

The final class of reference to physiology is the random mark or reference that, even if left uncompleted, provides further testimony as to the persistent influence of chemistry on Diderot's thought. An example of this is to be found in his fairly speculative discussion of respiration. Although this section is vague and contains more questions than answers, it does refer to Rouellian theories about the elasticity and compressibility of air. Diderot also postulates here that air mixes with blood in the process of respiration.[128] This, of course, would be one of Lavoisier's major breakthroughs, achieved in 1777.[129] Tantalisingly Jean Mayer considers the possibility that Diderot actually arrived at the idea before Lavoisier, and in any event holds that it was a remarkable idea to have had at the time.[130]

Further examples of cursory but potentially significant remarks occur in the second appendix of Mayer's critical edition of *Eléments de physiologie*. This appendix presents a collection of scattered notes that Diderot composed around the same time as the work proper, and which are more or less connected to it, at least in terms of subject matter. Here is mentioned a synthesis performed by Rouelle in which an alkali solution and air are combined to produce vitriolic tartar (potassium sulphate).[131] There is also a brief but intriguing discussion of electricity, in which Diderot ponders the possibility of collecting electricity 'by a process analogous to chemistry'. Unfortunately the comment is not taken any further here, and the reader is left rather hanging. Does Diderot mean that electricity itself might be decomposed? Or is he merely suggesting that instruments common in the chemist's laboratory might be useful in harnessing electricity?[132] In any case, Diderot applies Rouelle's lesson that principles themselves are not presented to our senses when he states that 'le feu électrique est un feu élémentaire très composé [i.e.: it is not pure, but rather a composite], car il a odeur et couleur'.[133]

Taking together the three orders of reference outlined above, *Éléments de physiologie* provides the richest and most rewarding hunting ground for proof of how Rouelle's chemistry enlightened Diderot's later thought. It

128. *Eléments de physiologie*, Quintili edn, p.213.

129. See Charles A. Culotta, *Respiration and the Lavoisier tradition: theory and modification, 1777-1850* (Philadelphia, PA, 1972).

130. See Mayer's note in *Eléments de physiologie*, Mayer edn, p.141.

131. *Eléments de physiologie*, Quintili edn, p.391.

132. The answer would seem to be suggested in a letter of 24 April 1773 from Mademoiselle d'Epinay to the abbé Galiani, in which the former relates a conversation with Diderot in which the *philosophe* is to have said that apparatus may be devised to decompose electricity and to combine its constituents (thought to be 'feu élémentaire' and various inflammables) with other substances. See Diderot's *Correspondance*, vol.12, p.206.

133. *Eléments de physiologie*, Quintili edn, p.391.

also raises many questions, such as the matter of exactly how Diderot's application of chemical concepts may have differed from that of his contemporaries. Was Diderot unique in his chemistry-inspired characterisation of various physiological elements, or is his method of description more or less the same as that which occurs in the work of Haller, Bordeu, Whytt, Marat or Le Camus? Certainly Haller attempted to grasp the composition of certain parts of the body, such as fibres, in terms of chemistry, but how close is his method of analysis and description to that of Diderot?[134]

Questions such as these require further research. What is apparent at present is that Rouelle's lectures exposed Diderot to a number of crucial ideas that were essential to the formulation of the *philosophe*'s later thought. But of course Diderot did not accept the chemistry of Rouelle *tout court*; certain aspects of Rouelle's chemical theory are conspicuously absent (or nearly absent) in the *philosophe*'s writings. There is in Diderot's work only one reference to phlogiston (*Réfutation d'Hemsterhuis*), and it is a reference that does not carry with it any connotations of belief in the 'inflammable principle'.[135] This probably indicates that, like Buffon, Diderot regarded the theory as a needless and unverifiable hypothesis. Likewise, Diderot seemed to reject that other Stahlian concept of the latus. Although Diderot concurred with Rouelle that extra-gravitational, extra-mechanical forces need to be considered in explaining the nature of chemical attraction and affinity, he never actually launched into any speculation as to what the basis of this might be (such as Rouelle's suggestion that it is the 'homogeneity of surfaces'). A final element of Rouelle's course that is notable in its absence from Diderot's work is alchemy. This is not necessarily to say that Diderot was therefore more modern or more forward-looking than Rouelle, but simply that he seems to have refused to enter into speculation as to whether or not alchemy is possible.

What then is to be made of Diderot's chemistry? Charles Coulston Gillispie once suggested that the chemistry (and indeed the entire natural philosophy) promulgated by Diderot and certain other of Rouelle's students, such as Venel, was essentially antithetical to the development of modern science. Seen to adhere to a pre-Newtonian (that is, Stahlian) concept of chemistry and a methodology completely enraptured with the idea of *pressentiment*, Diderot is portrayed by Gillispie as one hopelessly at odds with the rising tide of quantification and rationalisation, as the proponent of a romantic and humanistic but soon to be outdated approach to the study of nature. On the basis of this reading, Gillispie places Diderot at the centre of an *encyclopédiste* coterie that, he argues,

134. Thomas S. Hall. *Ideas of life and matter*, 2 vols (Chicago, IL, 1969), vol.2, p.62.
135. *Réfutation d'Hemsterhuis*, p.50.

exercised a profound influence on the Jacobin backlash against science; a backlash that resulted in the dissolution of the *Académie* and the execution of Lavoisier.[136] Such an analysis, however, is wide of the mark. First, as we have seen immediately above, several of the most crucial elements of Stahlian chemistry were virtually ignored by Diderot in his discussion of the subject; neither phlogiston, nor the latus theory nor alchemy ever made a significant impact upon his thought and work. Second, as the previous chapter has demonstrated, Diderot's methodology relied upon and emphasised much more than intuition – his concept of *pressentiment* did not preclude in any way the need to frame and to communicate theories and findings in a clear, rational and often even mathematical fashion. Moreover Diderot did not really see or use chemistry to spearhead a shift away from the Newtonian picture towards something more ambiguous and ultimately untenable. For him, the discipline was entirely consonant with a world-view that placed emphasis on order, structure, regularity and the presumption that things work in accordance with predictable, empirically verifiable laws.

Most significantly, however, we may object to Gillispie's appraisal on the basis that it is unduly whiggish. By viewing Diderot's chemistry with an eye primarily to what it lacks in the context of subsequent developments (and with the benefit of hindsight), attention is removed from the issue of what it actually is and the manner in which it served his particular natural philosophy. This, indeed, has been the purpose of the current chapter. Ultimately, it has not been our purpose to evaluate Diderot's work vis à vis Lavoisier or to consider his contributions to the theoretical development of modern chemistry, but simply to understand the manner in which Diderot learned about the discipline, the way in which it was constituted for him, and the effect that this had on his thought.

Having sketched these things out, a number of observations may now be added to Jean Mayer's assertion that Rouelle imparted to Diderot key concepts regarding the transformation of matter and the notion of vital chemical energy which enabled the *philosophe* to penetrate beyond a mechanistic view of the universe. First, an appreciation has been obtained as to exactly how this transferral of ideas occurred. The development of Diderot's thoughts on these issues has been related in a direct and substantive way to specific elements of the content of Rouelle's course. It has also been demonstrated that the impact of chemistry on Diderot's thought was more extensive than the provision of two or three essential, though very general, concepts. Ideas about scientific method, and about the need to broaden the accepted definition of matter, were given a

136. Charles Coulston Gillispie, 'The Encyclopédie and the Jacobin philosophy of science: a study in ideas and consequences', in *Critical problems in the history of science*, ed. Marshall Clagett (Madison, WI, 1962), p.255-89.

firmer, more positive basis on account of chemistry. The notions that a whole can be qualitatively greater than the sum of its parts, and that combinations of matter can and do lead to the emergence of novel, wholly unpredictable properties, were also seminal. Diderot learned from Rouelle that the kingdoms of matter are intimately linked and that through the process of decomposition and recomposition a given particle can participate in substances proper to any of the 'règnes'. Rouelle's course also helped Diderot to appreciate fully the utility of chemistry and even furnished the *philosophe*'s linguistic arsenal with new terms, analogies and metaphors. Diderot also gained from Rouelle a strong sense of how chemistry overlapped with both experimental physics and physiology, and how it could not only make important contributions to these disciplines, but form something of a bridge between them. With regard to physiology, chemistry was able to point the way towards an analysis of the human body that went well beyond mere form. In terms of the relationship between chemistry and physics, Diderot was able to employ Rouelle's lessons to posit a realm of operations that lay beyond but did not refute the Newtonian laws of movement. Given this, we might even qualify Mayer's assertion that chemistry served to bolster or propel Diderot's anti-mechanism. Certainly, Diderot opposed Cartesian single-force mechanism, but if mechanism is understood in more Newtonian terms, then the situation is far from clear. On the basis not only of this chapter but the previous one as well, it can be seen that Diderot did not seek to simply brush mechanism aside, but to redefine certain essential aspects of it, put it into perspective, contemplate it anew and demonstrate that more subtle, empirical, phenomenological modes of thought must be superadded to it in order to achieve a better grasp of the world. In other words, Rouelle's chemistry became thoroughly enmeshed for Diderot with other means of regarding nature; it sophisticated, deepened and nuanced in a profound manner his view of reality. Taken together, all of these ideas and effects allowed Diderot to develop a world-view that embraced not just transformism and the notion that molecules are animated by internal as well as external forces, but also qualitative change, diversity, dynamism and a perception of the universe as being almost infinitely complex and yet still fundamentally ordered. Moreover, information acquired from Rouelle's lessons allowed him to posit this world-view as part of an unprecedentedly integrated, scientific, well-argued position, rather than as fanciful speculation. Thus chemistry helped to create the conditions which made possible Diderot's mature natural philosophy.

4. Natural history

Considerations of Diderot's natural philosophy have long tended to centre around what may loosely be termed his 'biological' thought; that part of his work which concerns itself with the dynamic, living universe, and with the particular and individual qualities of the things that comprise reality. This is appropriate, or at least understandable, since frequently it appears that the *philosophe*'s fecund mind is working at its best when discussing ideas surrounding the origin, nature, forms and functions of life. Indeed, this branch of thought is the stage for some of Diderot's most daring and original ideas. It has also been perceived that in promoting the study of life in and of itself Diderot was, along with a small minority of eighteenth-century thinkers, disparaging traditional approaches to the study of nature. The idea here is that a key part of Diderot's philosophical credo involved displacing a view of nature based on physico-mathematics and replacing it with a more descriptive, experimental, 'natural historical' approach. A crucial aspect of this shift is the presumption that Diderot preferred to regard life itself as an autonomous principle or force, rather than reducing it to mechanical laws such as those which applied to physics. An extension of this idea is Aram Vartanian's contention that even early in his career, Diderot was on novel ground in acknowledging that the proofs and criteria of acceptability which applied to physics were different from those which applied to biology.[1]

While there is undoubtedly some truth in all of this, there is also danger in driving too thick a wedge between Diderot's interests in the organic world and the rest of his natural philosophical canon. Part of the problem with the appraisal of Diderot indicated above is that in emphasising the originality of his thoughts on life, the extent and profundity of his ideas regarding mathematics, mathematical physics, experimental physics and chemistry are misconstrued or underappreciated. As the previous chapters have revealed, Diderot's work in these areas was extensive, well considered and developed as part of an overall attempt to make sense of the universe. Ultimately, his views on the nature of life can be no different; they are likewise but a part of this grand framework, and must be comprehended in a fashion that links them with other areas of his thought.

1. Aram Vartanian, 'From deist to atheist: Diderot's philosophical orientation 1746-1749', in *Diderot studies* 1 (1949), p.48-49. The use of the term 'biology' here is, of course, anachronistic, as the term did not exist prior to 1800.

A compelling reason for interweaving or harmonising Diderot's thoughts on the 'physical sciences' with his ideas pertaining to the 'life sciences' is that to speak of his work in such stark terms is anachronistic. The boundaries between the study of inert matter and organic matter were not entirely clear in mid- to late-eighteenth-century France. Attempts to define life itself were bound up with heated philosophical debates as to whether or not living matter could be treated in a fashion different from other types of matter, and if so, how this might be achieved. This worked the other way as well, of course – philosophers who hoped to explain the origins, nature and functions of life in terms of inorganic or inanimate matter were faced with equally serious theoretical questions. Thus, if the 'life sciences' were emerging as a discipline somewhat independent of the other realms of natural philosophy in the mid eighteenth century, this trend was still very much in its infancy, and many interpretations of the relationship were still possible.

Even more importantly, at least with regard to Diderot studies, is that both aspects of natural philosophy were underwritten by and developed in accordance with many of the same presumptions in mind. These include notions about proper methodology, the relationship between an object and its interpreter, as well as the nature of matter, force and ultimately nothing less than the fundamental principles at work in the universe. In other words, it is not only possible but necessary (despite all temptations to the contrary) to understand Diderot's 'life science' in conjunction with and connection to his 'physical science'. Only in such a manner can the meaning, place and true originality of Diderot's natural philosophical world-view be thoroughly appreciated.

Such an endeavour must begin by situating the modern term 'life sciences' within the context of the eighteenth century, and more specifically within the context of Diderot's own thought. What was the nature and scope of his interest here? What constituted, for Diderot, what would now be called the 'life sciences'? There are two primary realms of study that, on a superficial level, subsume the bulk of Diderot's thought pertaining to the various phenomena of life. These two areas vary widely in terms of breadth and also, perhaps, in terms of their connection to the rest of Diderot's natural philosophy. One of these fields is medicine, which shall be the focus of the final chapter of this book. The other realm, which in the modern sense (though perhaps not in that of the eighteenth century) is much broader and in many ways more vaguely defined, is natural history.

Even within the scope of natural history it is essential to determine the precise extent and emphasis of Diderot's interest, for his involvement in discussing issues pertaining to natural history was by no means spread equally throughout all aspects of this massive branch of learning. By the middle of the eighteenth century, the pursuit of natural history in France, and throughout Europe, was developing along two radically different lines. One school of thought emphasised a slow, cautious, discreet, highly empirical approach to nature. It was concerned primarily with collecting specimens, describing them accurately and, if possible, orienting them within some system of classification. This approach has been connected with the attitude to natural philosophy of Francis Bacon, who advocated the compilation of extensive 'natural histories' upon which to base any attempt at induction.[2] It was also very notably linked to the development of natural theology, a branch of learning which sought to demonstrate the existence of a beneficent Creator by reference to the order and marvels of His grand design.[3]

Perhaps the greatest exemplar of this mode of natural history is the Swede Carolus Linnaeus (1707-1778). A botanist by training and founder of the Royal Swedish Academy of Sciences (in 1739), Linneaus developed as the foundation of all his work a system for the classification of plants. This binomial system (using only two names to classify each plant – one referring to its genus and the other to its species) was enormously influential and won Linnaeus international acclaim. He made full use of his reputation and his power, organising expeditions to various corners of the world in order to take further stock of nature's bounty, seeking in each instance to describe it well and then fit it into his classificatory system. Linnaeus was largely uninterested in the history of living forms (although he did make a limited attempt at some such understanding, largely, it should be noted, in terms of Mosaical history), his grand view, such as it was, tended rather to simply emphasise the harmony and balance of nature's economy. In other words, his view of the natural world was very much aligned with that of the natural theologians; it was perceived as a fixed, stable, finite system.[4]

In France, such a tradition was best represented by R. A. Réaumur (1683-1757), who held a prominent position within the Académie des sciences. Probably a more versatile intellect than Linnaeus, Réaumur possessed interests encompassing the natural philosophical spectrum, including mathematics, electricity and the practical applications of contemporary science. In connection with the latter item, he was supervisor of the Académie's industrial encyclopedia project, a role that may be seen as having made him an instant rival of Diderot. Most notably, however, Réaumur cast a long shadow over the world of eighteenth-century natural history, having arguably dominated the discipline in France from the 1710s. He was primarily interested in insectology, and between 1734 and 1742 published six volumes on invertebrates.

2. Phillip Sloan, 'Natural history, 1670-1802', in *Companion to the history of modern science*, ed. R. C. Olby *et al.* (London, 1990), p.296.

3. Sloan, 'Natural history', p.297.

4. On Linnaeus see A. Stafleu, *Linnaeus and the Linnaeans: the spreading of their ideas in systematic botany, 1735-1789* (Utrecht, 1971), and Tore Frängsmyr, *Linnaeus, the man and his work* (Canton, MA, 1994).

The means of study advocated by men such as Linnaeus and Réaumur dominated natural history during the first half of the eighteenth century and, to a large degree, much of the second half as well. Although the precise methods and activities of the two men may have differed slightly, Réaumur, for instance, was much more interested in observation than classification,[5] they both concerned themselves primarily with investigating the structures of natural objects. The essence of their approach was the description of nature as it was laid out before them. While such a task was by no means simple (indeed, valid factual observations were presumed to require a properly trained eye and good sense of judgement), it did give the science of natural history an accessibility that was lacking for most other branches of natural philosophy. The eighteenth century was, after all, the great age of natural history cabinets, of the collection and classification of specimens, of indulgence in a pastime that was open to contributions from amateurs and professionals alike.[6] Among the *philosophes*, Jean-Jacques Rousseau is an ideal example of this trend. Although he was not fond of botany as a science, he was a passionate collector of specimens, often spending days outside the cities picking flowers, identifying plants and, in general, appreciating the bounty and diversity of God's glorious production.

But if this traditional, orthodox, and relatively straightforward approach to natural history dominated for much of the eighteenth century – to the extent that Michel Foucault once claimed that the century's overemphasis on simple observation and classification of the natural world made the true emergence of 'biology' impossible[7] – it was not the only means of proceeding. All naturalists were reliant upon the assiduous collection of facts and close observation of nature, but some attempted to go well beyond the parameters sketched by Réaumur and Linnaeus in terms of seeking to understand them. Foremost in this respect was George Louis LeClerc, comte de Buffon, who was *intendent* of the Jardin du Roi (from 1739 until his death in 1788)[8] and author of the encyclopedic *Histoire naturelle* (published from 1749-1789). In the *Discours préliminaire* to his *Histoire naturelle* Buffon set up a programme of natural history that was daunting in both scope and ambition. In laying out his

5. Roger, *Buffon*, p.71. Elsewhere, Roger emphasises that harping too much on the preponderance of classification in eighteenth century-natural history does the discipline a disservice, for there was much going on simply in terms of observation and description. See Jacques Roger's article in G. S. Rousseau and Roy Porter, *The Ferment of knowledge* (Cambridge, 1980), p.265. This point is also well demonstrated by Laurence Brockliss in *Calvet's web: Enlightenment and the Republic of Letters in eighteenth-century France* (Oxford, 2002), p.267-70.

6. A useful study of collecting in Britain is provided by Patricia Kell, *British collecting 1656-1800: scientific enquiry and social practice* (D.Phil thesis, Oxford, 1996).

7. See Michel Foucault, *Archaeology of knowledge* (London, 1972).

8. For more on Buffon's role as head of this significant institution, as well as an account of its flowering under the nearly fifty years of his administration, see Spary, *Utopia's garden.*

goals, Buffon made the seminal distinction between particular natural histories and general natural histories. The former referred to the slow, patient process of observing nature, collecting facts and remaining satisfied with their enumeration and classification; that is, with describing and systematising nature's manifold productions. The description of the structure of a particular species and its ascription to a certain genus would be an example of a particular natural history. Assembling a general natural history, by contrast, involved trying to achieve a deeper understanding of nature and its processes. Buffon himself believed, in very Lockean fashion, that after looking long and diligently at the varied phenomena of nature, certain facts and ideas start to impress themselves upon us and subsequently to 'connect themselves in our spirit by means of fixed and invariable relations; and from these we elevate to more general views'.[9] As Jacques Roger has described it, Buffon's aim was nothing less than to gain access to the 'engine room' of nature, that is, penetrate beyond the mere appearance of things and reduce nature's operations to systems, thereby grasping the 'big picture'.[10]

Buffon announced in the *Discours préliminaire* that his natural history would be both particular and general; indeed the proper title of his masterpiece is *Histoire naturelle générale et particulière*. This proclamation of intent seemed audacious to many contemporaries, some of whom openly scoffed at the project. Réaumur, for example, argued that so much remained to be done in the way of assembling particular natural histories that no one could even dream of commencing a general one.[11] Nevertheless, Buffon persevered, and went on to produce a work that encompassed both description of nature and discussion of its underlying physics.[12] This bold stroke involved more than just a change of goals, it represented a wholesale shift in attitudes towards nature. Where the work of men such as Réaumur emphasised the stability and fixity of natural phenomenon, as though the immensity of the natural world was simply laid out before Man in a purely spatial dimension, Buffon's natural history was underwritten by the assumption that temporal change was a fundamental facet of the universe. Rather than presuming the existence of an immutable natural system, Buffon's work was predicated upon a sense of dynamism, and sought to come to terms with relationships and processes. Thus, his *Histoire*

9. Buffon, *Histoire naturelle*, taken from Spary, *Utopia's garden*, p.195.

10. Roger, *Buffon*, p.75.

11. Roger, *Buffon*, p.75.

12. Roger claims that in this, Buffon was influenced by Stephen Hales, who had attempted to introduce Newtonian physics into natural history by discussing the roles of heat, fermentation and attraction as fundamental forces of life, forces responsible for creating and guiding the life processes of plants and animals. See Roger, *Buffon*, p.26. See also Stephen Hales, *Vegetable staticks*, foreword by M. A. Hoskin (London, 1961, first published in 1727). Chapter 3 discussed how Rouelle was likewise impressed by the work of Hales, particularly the latter's physico-chemical attempts to unravel the nature of plant life (See chapter 3, p.101 and n.42).

naturelle dealt to an unparalleled degree with issues such as the distribution of species, the migration of various forms and geological change.[13] Contrary to the prevailing practice of zoology, Buffon did not become bogged down in addressing the morphology of animals, but occupied himself with matters of generation, instincts, habits, diet, behaviour and environment.[14] Crucially, Buffon even considered the possibility of changes within species, suggesting that certain types of animals (particularly smaller ones) may become weaker in unfavourable environments, creating the situation whereby some varieties of animal might be regarded as degenerations of a prototype.[15] In other words, he promoted the notion that species themselves have a history.

Significantly, the gap that existed between Réaumur and Buffon was not simply intellectual. Their rivalry was coloured by profound personal enmity, lending the dichotomy between two competing strands of natural history a highly political dimension. Within France at the mid-point of the eighteenth century two distinct camps of naturalists had arisen; one centred around Réaumur and the Académie des sciences, the other around Buffon and the Jardin du Roi. Initially, Réaumur's position was by far the stronger. As has been mentioned, the ageing academician had held sway over the world of French natural history from the early part of the century. He was at the centre of a rather remarkable international correspondence network, and counted amongst his allies renowned scientists such as Charles Bonnet (1720-1793) and Abraham Trembley (1710-1784).[16] By the century's end however, it was the approach adopted by Buffon that had become prominent, and despite the fact that his *Histoire naturelle* had met with a very critical reception from members of the Réaumur circle upon its appearance (Bonnet, for instance, lambasted it as unscientific and artificial), Buffon's masterpiece went on to become one of the best selling books of the century.

The place occupied by Diderot within this shifting and somewhat treacherous landscape was unambiguous: he was firmly on the side of Buffon, who, as shall be seen throughout this chapter, exerted a profound influence on the *philosophe*'s thoughts about the natural world.

13. Sloan, 'Natural history', p.305.

14. Roger, *Buffon*, p.89. It should be noted that in addressing these matters, Buffon was going well beyond the Newtonian platform presented by Hales.

15. Antonello Gerbi, *The Dispute of the new world; the history of a polemic, 1750-1900*, translated by Jeremy Moyle (Pittsburgh, PA, 1973), p.22-33.

16. Spary, *Utopia's garden*, p.19. Charles Bonnet was a Swiss naturalist and philosopher who received widespread acclaim for his work *Contemplation de la nature* (1764). He was a prominent figure in eighteenth-century embryological debates, maintaining, like Réaumur, a preformationist position (for more on preformationism, see chapter 5). Abraham Trembley was another Swiss, best known for his work on the regenerating freshwater polyp, or hydra (for more on this, see below, p.137-38). The most notable works on Bonnet and Trembley are Virginia Dawson, *Nature's enigma: the problem of the polyp in the letters of Bonnet, Trembley and Réaumur* (Philadelphia, PA, 1987), and Lorin Anderson, *Charles Bonnet and the order of the known* (Boston, MA, 1982).

Unsurprisingly then, Diderot was generally unimpressed by the approach to natural history represented by men such as Linnaeus and Réaumur. As such, what Buffon referred to as particular natural history is conspicuous in its absence from Diderot's canon of work. His 'natural history' is immediately marked as being strikingly different from that of many of his contemporaries. It must be granted of course that he was not totally ignorant of the advantages of pursuing natural history along what might be termed 'traditional' lines. *Plan d'une université* speaks rather highly of such classic pursuits as collecting and describing, proclaiming that the science is extremely useful and interesting, and that it is well suited to the education of young children, as it is 'un exercise continu des yeux, de l'odorat, du gout et de la mémoire'.[17] It should also be observed that the method of Réaumur was undoubtedly affiliated with certain Baconian elements of *Pensées sur l'interprétation de la nature* (in terms of its enthusiasm for natural philosophy that is based on patient, careful observation). Nevertheless, Diderot himself showed little inclination towards taking up conventional activities such as botany. Part of the reason for this may simply be due to the circumstances of his life – he was confined, as it were, within the city of Paris for much of his adult life. His twenty-five-year labour on the *Encyclopédie* was intense, often hectic and afforded him little time to stray far from the editorial desk.[18] Nevertheless, his lack of participation in one of the century's most rapidly growing and popular pastimes was certainly down to choice as well. For one, he was a self-professed city-lover, having become addicted to the bustle and activity of Paris almost from the day he arrived.[19] Second, opportunities to pursue activities like botany did arise, particularly later in his life. Occasional stays in the country at Grandval, the home of Baron d'Holbach, or even his grand tour through Holland, Germany and ultimately to Russia would have afforded him the opportunity to give botany a try if he had been so inclined. But he was not.

More importantly, not only was Diderot uninterested in pursuing botany as a 'sport', but his writings display little concern for discussing the practice of botany, zoology, geology, or any other branch of what is typically perceived to constitute 'natural history'. In fact, Diderot was not only bored by the prospects of following or writing about such pursuits, but he actually held them in some measure of disdain. Evidence of this was his criticism, to be discussed in greater detail shortly, of Linneaus and the Linnean system of classification. Diderot had strong philosophical objections to the entire project and purpose of the Swedish naturalist, and

17. *Plan d'une université*, p.779.

18. A good example of the potential time, effort and expense required to pursue natural history – even as a sideline – is indicated in Brockliss, *Calvet*, p.249, 251-52 and 254-63.

19. Crocker, *Diderot: the embattled philosopher*, p.27.

argued in *L'Interprétation de la nature* that logic itself speaks against his methods.[20]

The realm of zoology also came under fire from Diderot. *Pensée 54* of *L'Interprétation de la nature* mocks Réaumur, and belittles the value of the latter's work on insectology:

> Les hommes extraordinaires par leurs talents se doivent respecter eux-memes et la postérité dans l'emploi de leur temps. Que penserait-elle de nous, si nous n'avions à lui transmettre qu'une insectologie complète, qu'une histoire immense d'animaux microscopiques? Aux grands génies, les grands objets; les petites objets, aux petites génies. Il vaut autant que ceux-ci s'en occupent, que de ne rien faire.[21]

As Aram Vartanian has noted, Diderot seems here to 'dismiss as somehow secondary, even trivial' the acheivements of naturalists in the Réaumur mold.[22]

Part of Diderot's virulent attitude may have stemmed from a long-standing antipathy between the two men, which could have been a function of Diderot's affiliation with the 'Buffon camp' of naturalists.[23] Demonstration of this ill feeling is revealed by an incident that took place in 1748. Around the time Diderot was writing *Lettre sur les aveugles*, Réaumur conducted an operation which involved removing a cataract from a person who had been blind from birth and observing their immediate reaction. Obviously, this experiment was of profound interest to Diderot, and he sent a letter to Réaumur formally requesting the privilege of attendance at the moment that the subject's bandages were removed. Réaumur initially declined the request, and when, after some persistence on Diderot's part, he finally gave in, he invited the *philosophe* to what was effectively a false trial; the ceremonious removal of bandages that Diderot witnessed was not in fact the first time they had been taken off. When Diderot discovered this he was enraged, realising that he had in fact failed to witness the true experiment altogether. It has even be speculated that Réaumur and/or his friend Mme de Sainte-Maur saw to it that Diderot's impudent protests were punished by a spell in the prison at Vincennes. It should be noted however that Jacques Roger tempers the validity of this intriguing piece of gossip by noting that even before the incident Diderot had been marked by the police as a dangerous author.[24]

20. *Pensées sur l'interprétation de la nature*, p.755.

21. *Pensées sur l'interprétation de la nature*, p.762.

22. Vartanian, *Science and humanism in the French Enlightenment*, p.114.

23. This, of course, would not be the only instance of Diderot attacking the nature of a contemporary's preferred discipline at least partially on the basis of personal differences. We have seen in chapter 1 that he made a similar move with regard to his erstwhile friend D'Alembert and the discipline of pure mathematics.

24. Franco Venturi, *La jeunesse de Diderot, 1713-1753*, translated by Juliette Bertrand (Paris, 1939), p.186-87. It should also be noted, of course, that as an orthodox Christian, Réaumur was likely quite to have been offended by Diderot's *Lettre sur les aveugles*.

In any case, the main issue is that Diderot did go out of his way to criticise the value of Réaumur's professional pursuits. Moreover, his attack was not simply a personal indictment, but, as we have seen, was widely applied to the entire field of insectology. Both botany and zoology appear therefore to be veritably condemned by Diderot. Given this, we must ask the obvious question: what was Diderot's interest in 'natural history'? Simply put, Diderot was inclined towards what Buffon had termed general natural history. Although disparaging of the practice of natural history in its conventional and popular eighteenth-century sense, Diderot was tremendously interested in the theoretical considerations which both underlay and arose from it. He was, we might say, not so much interested in natural history as in the history of nature. He was a cosmologist, occupied with questions pertaining to the nature of life, and the true nature and history of the world. If he was largely uninterested in the raw material of 'natural history', he was consumed by many of the questions which were drawn from it, or relied upon its evidence. The key to Diderot's interest here was not to describe the forms of nature's productions but to understand them, and upon this basis to consider all relevant philosophical consequences and implications. If we are to ascribe to Diderot an interest in something called 'natural history', then the discipline must be understood in a very broad sense of the term. By all means Diderot believed that any attempt at understanding must be rooted in solid, careful observation of nature, but ultimately he was most interested in interpreting these observations, and assembling them into some kind of coherent world-view.

For Diderot this process involved coming to terms with some of the eighteenth century's most heated debates. One of these was stimulated by Abraham Trembley's discovery, in 1744, of the regenerating freshwater hydra, or polyp. Trembley found that each piece of a dissected specimen would, in time, regenerate to form a complete polyp. This capacity confounded many philosophers and theologians, who found that traditional notions of the soul, or at least of the relationship between the life of a being and its organisation, were severely challenged by this curious creature. If each part of a polyp was capable of reforming itself into a whole polyp, where was the centre of its being located? Where or what was its 'soul'? Such considerations were of much more than passing or purely theological interest. Even for those who may not have ascribed 'souls' to animals – much less to those as base as the polyp – the very existence of such a bizarre creature left many significant and perplexing questions.[25] Was the principle of its life, of its very existence disseminated

25. The issue of animal 'souls' was, of course, a very important one to many early modern thinkers. Although Descartes had denied that animals had souls, and later La Mettrie would deny the validity and applicability of the word 'soul' even to human beings, there were certainly those who persisted in ascribing souls to all manner of animals. See, for

throughout every particle of the animal? If it was, how could this be so? How did this function mechanically? And what consequences did this have on perceptions about the nature and operation of life in general?

Similar controversy surrounded the supposed discovery of spontaneously generated life by John Turberville Needham (1713-1781). Having isolated inert matter in a vessel, Needham observed the contents with a microscope, and after some time identified a profusion of tiny 'eels' swimming about. This he took to be evidence of spontaneous generation – the idea that life can spring spontaneously and naturally from non-living matter. Needham's 'discovery' had a great many detractors, not least among them Voltaire, who wholeheartedly resisted Needham's conclusion, even becoming involved in a protracted debate with the microscopist about the possibility of spontaneous generation.[26] Subsequent developments proved Voltaire's protestations to be correct, though not on account of his fervently held deist world-view. The veracity of Needham's conclusion was undermined rather by careless experimental procedure; he had failed to properly seal the vessel, thereby allowing air, and with it airborne micro-organisms, into the jar. Similar experiments conducted later in the century by Lazzaro Spallanzani (1729-1799) used perfectly sealed vessels, and no signs of spontaneously generated life were detected.[27] Nevertheless, the Needham experiment made an impact at the time. It stirred up controversy, and for much of the century was upheld as a demonstration of the possibility that life might simply spring from inert matter.[28]

These ideas, to be taken up in the context of Diderot's own thought shortly, were intimately connected to and subsumed under the larger question of what life is, and how living matter is distinguished from seemingly dead or 'inert' matter. In tackling these issues, Diderot relied heavily on the published memoirs of the Académie des sciences, in which he learned about the work and views of natural philosophers including Trembley and Needham.[29] An even bigger source of information for Diderot however was the work of Buffon. Jacques Roger credits Buffon

instance Antoine Dilly, *Traite de l'ame* (Paris, Gallet, 1691), and David-Renauld Boullier, *Essais philosophiques* (Amsterdam, 1728). For more on eighteenth-century debates about the existence and prospective nature of animal souls, see Jacques Roger, *The Life sciences in eighteenth-century French thought*, ed. Keith R. Benson, translated by Robert Ellrich (Stanford, CA, 1997).

26. For more on this debate see Shirley Roe in John W. Yolton (ed.), *Philosophy, religion and science in the seventeenth and eighteenth centuries* (Rochester, NY, 1990), p.418-19.

27. Lazzaro Spallanzani, *Nouvelles recherches sur les découvertes microscopiques, et la génération des corps organisés. Ouvrage tr. par l'abbé Regley. Avec des notes, des recherches physiques & métaphysiques sur la nature & la religion [&c.] par m. de Needham* (London, 1769).

28. See Roger, *The Life sciences in eighteenth-century French thought*, p.486-88

29. Mayer, *Diderot, homme de science*, p.54.

with having brought to Diderot's attention many of the most significant aspects of eighteenth-century thought regarding the nature of life.[30] Diderot was first exposed to the work of Buffon during his imprisonment at Vincennes in 1749. It has been tantalisingly suggested that during this time he composed an extensive commentary to the *Discours préliminaire* of Buffon's *Histoire naturelle*.[31] Unfortunately this work is no longer in existence, having presumably been destroyed at some point during Diderot's confinement. In any case, there is little doubt that Diderot was tremendously impressed by Buffon and his work. Throughout many of Diderot's writings there is frequent praise for the *Histoire naturelle*, its author and his ideas. Further testimony regarding the extent of Diderot's appreciation for Buffon's work is that he often borrowed entire passages for insertion in the *Encyclopédie*. Perhaps the most famous example of this is the article 'Animal'. Much of the substance of this piece is taken directly from *Histoire naturelle*, with most of the remainder serving as little more than a linkage of or commentary on Buffon's thought.

Jacques Roger has posited that Diderot and Buffon also enjoyed a close personal relationship. There is little concrete evidence lending weight to this supposition, the correspondences of both men, for instance, containing relatively few letters addressed to the other. Nevertheless, that Diderot and Buffon may have known each other personally and been well disposed towards one another is far from inconceivable. Roger claims, for instance, that while in prison Diderot had Buffon vouch for his character, and this claim is backed up by Franco Venturi's work on the youth of Diderot.[32] Even during the later years of the publication of *Histoire naturelle*, when most were greeting it with less enthusiasm than before, Diderot continued to lend the work his unflagging praise and support.[33] Just as significantly, Buffon seems to have possessed the same high regard for Diderot's work, volume three of *Histoire naturelle* recommending to its readers *Lettre sur les aveugles*, declaring it to be a work of 'metaphysique très fine et très vraie'.[34] These episodes certainly appear to involve acts of genuine friendship, or at least profound personal respect, rather than mere professional acquaintance.

In any case, it is clear that Buffon played a leading role in increasing Diderot's knowledge of the key issues involved with the life sciences, inspiring much of his thought in this area and providing him with a plethora of information to ponder. This included ideas regarding the age, structure and history of the earth, the nature of animal species and the various forms of animal behaviour, the nature of the distinction between

30. See Jacques Roger, 'Diderot et Buffon en 1749', in *Diderot studies* 4 (1963), p.222.
31. Roger, 'Diderot et Buffon', p.222.
32. Venturi, *La jeunesse de Diderot*, p.187.
33. Roger, *Buffon*, p.199.
34. Roger, 'Diderot et Buffon', p.222.

the organic and inorganic realms of matter, and of course, as Diderot's presence within the 'Buffon camp' of natural historians would indicate, the proper means for studying and attempting to understand the functions of plants and animals. We can also see the influence of Buffon upon Diderot's aforementioned critique of Linnaeus' system of classification. As we have seen in chapter 1, the two men held very similar views on the necessary limitations of mathematics in the interpretation of nature and on the qualifications that must be attached to any type of abstract thinking.[35] The influence of Buffon upon Diderot might even be witnessed in the arrangement of *Plan d'une université*. Diderot stated: 'Je classerai les sciences et les études, comme notre historien naturaliste, M. de Buffon, a classé les animaux [...] Il a parlé d'abord du bœuf, l'animal qu'il nous importe le plus de bien connaître.'[36]

To list these examples is merely to scratch the surface of the manifold ideas that Buffon placed before the eager mind of Diderot. Nevertheless, the great naturalist was not Diderot's sole mentor or supplier of information with regard to issues pertaining to the nature of life. Diderot read very widely, and the canon he covered is known to include works by Bonnet, La Mettrie, Robinet, Réaumur, Albrecht von Haller, Le Roy and de Maillet.[37] Perhaps the most important figure aside from Buffon in terms of the development of Diderot's ideas regarding the nature and history of life however was Maupertuis, whose relevance to the unfolding of Diderot's natural philosophy has already been touched upon in chapter 2. One of the most notable aspects of Maupertuis's thought was what Max Wartofsky termed the 'materialisation' of Leibniz's monad; that is, taking the German philosopher's concept of a self-contained, metaphysical principle of perpetual and dynamic identity, and making it an aspect of the material world. Maupertuis stressed that the universe was predicated upon change and the continual shifting of various forces; for

35. Part of Diderot's receptivity to Buffon's ideas undoubtedly had to do with shared methodological ideals. Unlike the case of Diderot being open to the ideas of Rouelle on account of shared methodological preferences however, Diderot was exposed to the thought of Buffon before the composition of *L'Interprétation de la nature*. This creates the scope for the methodology sketched out in that work to have been inspired, at least in part, by Buffon. For more on the dialogue taking place at this time between not only Diderot and Buffon, but Rousseau as well, see *Rousseau and the eighteenth century*, ed. M. Hobson, J. Leigh and R. Wokler (Oxford, 1992), p.229-38.

36. *Plan d'une université*, p.760.

37. *Eléments de physiologie*, Mayer edn, p.xiii-xiv. On Julien Offray de La Mettrie (1709-1751) see chapter 5 p.198. On Jean-Baptiste-René Robinet (1735-1820) see below, p.161. Regarding Albrecht von Haller (1708-1777), one of the eighteenth century's most renowned physiologists, see Shirley Roe, *Matter, life, and generation: eighteenth-century embryology and the Haller–Wolff debate* (Cambridge, 1981). The *encyclopédiste* Charles-Georges Le Roy (1726-1779) was Superintendent of the Royal Chase at Versailles and author of *Mélanges de physique et de médecine* (1771). Benoît de Maillet (1656-1738) penned the infamous *Telliamed*, a controversial work pertaining to the formation of the Earth and the origins of humankind.

him, process was seen to inhere in matter itself.[38] For this, some have credited Maupertuis with having renewed the possibility in France of coming to terms with the true nature of life and of achieving an understanding of living forms.[39] If this were the case, it could be argued that Maupertuis paved the way even for the work of Buffon. Maupertuis also introduced important ideas about heredity and launched the proposition that all animal life may ultimately be derived from a single 'prototype'. All of these concepts would become highly relevant to Diderot's thought. Indeed, with the possible exception of ideas surrounding the concept of heredity, every one of the notions listed above found its way into *L'Interprétation de la nature*. Perhaps the most famous example of this influence is *Pensée 50*, which devotes itself to a discussion of Maupertuis's theory of matter. Crucially, Diderot agreed with Maupertuis that the accepted definition of matter needed to be broadened, primarily to take account of the phenomena of life. The details of this discussion shall be addressed shortly, the point here is that the thought of Maupertuis was a seminal part of the background against which Diderot's ideas took shape.

Of course, the provision of ideas by contemporary natural philosophers is something very different from the development of Diderot's thoughts on the nature of life in their own right. As Roger has observed: 'Diderot was not satisfied to combine Buffon's vocabulary and Maupertuis's thought.'[40] It has been pointed out, for example, that the period between 1749 and 1753 was the zenith of Buffon's impact upon Diderot's philosophy.[41] After this it seems that Diderot, while still engaged in activities such as the reading and lending of moral support to *Histoire naturelle*, was moving on and developing ideas along a more independent line. The fruit of this development was the assembly of one of the eighteenth century's most coherent and original theories regarding the nature of life.

Part of the reason for Diderot's increased boldness and originality in discussing these themes from 1753 onwards may have been a burgeoning sense of confidence, a sense that he had acquired enough background knowledge of both contemporary theories and the evidence supporting them to begin addressing issues of life and its nature less as a student and more as a natural philosopher in his own right. Another key factor,

38. Max Wartofsky, 'Diderot and the development of materialist monism', in *Diderot studies* 2 (1952), p.291. A slightly tempered view of the Leibnizian influence on Maupertuis's thought is offered by Mary Terrall, who argues that Maupertuis incorporated certain elements of Leibnizian philosophy into his view of matter and his theory of life, particularly in the language of perception and appetition. Terrall points out, however, that Maupertuis differed substantially from Leibniz; being both less dogmatic and mechanical than the German, and rejecting the notion of a separate metaphysical plane existing parallel to the physical world. See Mary Terrall, *The Man who flattened the earth: Maupertuis and the sciences in the Enlightenment* (Chicago, IL, 2002), p.333-34.

39. Cassirer, *Philosophy of the Enlightenment*, p.392.

40. Roger, *Buffon*, p.340.

41. Roger, *Buffon*, p.340.

however, is that in enunciating ideas about the nature of life, Diderot was able to discuss ideas more proper to the other scientific disciplines, such as experimental physics. As we have seen, experimental physics was a subject that Diderot felt eminently comfortable discussing, and regarding which he had much to say in terms of both content and methodology. An impressive knowledge of contemporary chemistry would likewise wind up enlightening his ideas about life from the mid 1750s onward.[42]

This is to say that in formulating his ideas about life, Diderot relied not just upon those theorists whose work was most obviously relevant to such concerns, but also very heavily upon his own increasingly well-articulated views about what may be termed the physical sciences. To put it more succinctly, many of the fundamental views and principles delineated in the first three chapters of this book apply to this chapter as well. At the very least, it must be said that Diderot's 'natural historical' theories were developed in conjunction with and parallel to his thoughts on the other disciplines. All are part of the same framework, and are a result of the same basic drive: Diderot's intention to make sense of the world and to develop a coherent, penetrating world-view.

One of the most obvious examples of congruence between Diderot's 'physical science' and his 'natural history' is the relation between his critique of mathematics and his very similar suspicion of classification. These two cornerstones of his natural philosophy arise from a common root: his insistence that the development of ideas through the imperfect yet nonetheless very necessary process of abstraction inevitably separates the thinker from reality. Little more needs to be said about his attack on mathematics – that has been considered in chapter 1. What is of interest here is how Diderot's condemnation of Linnean-style systems of classification shares precisely the same concerns and follows the same lines as his critique of mathematics.

In the article 'Animal', Diderot argues that the boundaries between species, and indeed between the very kingdoms of matter, are ultimately ambiguous. Transitions take place by 'imperceptible degrees', making the task of determining where one set ends and another begins literally impossible. The best that the naturalist can do is to assign general definitions, but these, as a consummately human act of representing what actually exists as opposed to capturing its essence, is inevitably bound to remain an approximation. Thus Diderot warns against relying

42. The previous chapter has indicated that chemistry was indispensable in Diderot's elucidation of concepts such as *sensibilité universelle*. We have also seen that Rouelle's chemistry impacted directly upon such natural historical issues as the composition and linkage of the kingdoms of matter. It might also be noted that the rising importance of chemistry in Diderot's thought lent further weight to his aforementioned disparagement of subjects such as botany, which was regarded as inferior largely because it dealt only with appearances, and did not attempt to come to terms with the true nature of objects (as chemistry presumably did).

too heavily upon these approximations and resists the notion that such ideas might be taken as an expression of truth. Emphasising the point, Diderot noted that: 'Plus on examine la nature, plus on se convainc que pour s'exprimer exactement, il faudrait presque autant de dénominations différentes qu'il y a d'individus.'[43]

It follows, of course, that if every entity in nature were assigned its own definition then the universe would remain incomprehensible to Man. However, this is not what Diderot was trying to achieve. He merely wanted to place the natural philosopher's use of definitions into perspective. Employing statements that clearly resemble his views on mathematics, Diderot argued in the article 'Animal' that Man abstracts from the particular beings of nature ever more general ideas (in order to assemble a great quantity of objects within the scope of one point of view), but that since these ideas have been derived using methods of human invention they are entirely artificial. He goes on to remark that 'ces lignes de séparation n'existent point dans la nature'.[44] Thus his goal was to qualify the naturalists use of abstract thought, and to set limits on the extent and applicability of classification.[45]

Diderot sought to establish an approach to nature that went beyond describing and classifying the forms of objects and attempted instead to understand their functions. In *L'Interprétation de la nature* he lambasted the hollowness of the Linnean pursuit on the basis that there is not much use in designating if no effort is made to know the thing one designates.[46] Those who proceed in ignorance of this maxim Diderot pejoratively labelled 'méthodistes'.[47] Like geometers, 'méthodistes' were, for Diderot, those who put too much stock in the innate value of their own systems of thought, and in so doing remained blind to much of reality. This argument is one of the primary themes of *L'Interprétation de la nature*, and it is in a statement pertaining to natural history that Diderot gave this theme what may be its boldest and clearest expression; he disparaged the 'méthodiste' because 'au lieu de reformer ses notions sur les êtres, il semble qu'on prenne à tache de modeler les êtres sur ses notions'.[48]

Thus Diderot's natural history is grounded in the same epistemology that grounds his mathematics, or physics or even chemistry. All of these

43. *Encyclopédie* article 'Animal', p.66-67.

44. 'Animal', p.69. Note that these lines of separation, which don't really exist in nature, are just like the lines of the geometer (see *Apologie de l'abbé de Prades* and chapter 1).

45. This might be understood as applying both to taxonomists who acknowledged that their systems were artificial, such as Linnaeus, and those who attempted to establish or discover 'natural' systems of classifications, such as Antoine Laurent de Jussieu (1748-1836). On Jussieu see Peter Stevens, *The development of biological systematics: Antoine-Laurent de Jussieu, nature, and the natural system* (New York, 1994).

46. *Pensées sur l'interprétation de la nature*, p.719-20.

47. *Pensées sur l'interprétation de la nature*, p.754.

48. *Pensées sur l'interprétation de la nature*, p.754.

pursuits are ultimately qualified by the same considerations. The touchstone for Diderot's work in each realm is Man – a being that is relatively unimportant in the grand scale of the cosmos and yet which is the sole creator of meaning in the universe. While Diderot was a realist, positing the existence of objects outside of the human understanding, he vigorously asserted that any understanding of these objects must necessarily be related in terms of human (namely artificial and approximate) ideas. His approach to theorising about life is therefore inseparable from the approach he used in other disciplines. In Diderot's philosophy there are a number of basic methodological guidelines and preferences which impact upon all types of thinking about nature.

The first of these is that there can be no discussion of absolutes. For Diderot, such a concept was both impractical and unnecessary. As has been seen with regard to his physics, Man could never lay claim to the essence of reality and was never in a position to formulate any law or posit any truth in unequivocal, absolute terms. Nothing could be viewed as independent of everything else. In Diderot's universe all things are related, and therefore discussion of any given phenomenon must be relative and qualified.

Likewise, natural historical concepts and theories were to remain for Diderot eminently flexible. In a sense, the flexibility of natural philosophy was a manifestation of Diderot's ban on the employment of absolutes. In this branch of his natural philosophy, as in all others, knowledge was to be regarded as transient. Science never reaches the goal of complete understanding, but is always susceptible to revision and improvement.[49] Thus, a concept or theory once taken to be virtually impregnable may eventually be modified, or even placed in the service of an idea that was originally unrelated. Diderot asserted very firmly that the natural philosopher must always be aware of this, and that he must never remain too tied to a particular way of seeing things. Dogmatism was, for Diderot, one of the greatest enemies of good science.

Diderot was also, in natural history as in every other branch of science, committed to promoting the need for careful observation and experiment. *L'Interprétation de la nature* and many other works make this point forcefully, harping on the need for the natural philosopher to proceed methodically and with consummate attention to detail. Diderot realised

49. Certain philosophers might maintain that if science could never lay claim to the 'truth', then it could in no sense be described as improving either. Nevertheless, we might still cite criteria such as predictive accuracy to argue that while one theory may not be any more 'true' than another (at least in absolute, ontological terms), that it is 'better', or 'more useful'. Another possibility, which is close to the position maintained by Diderot, is that the development of science is asymptotic to the truth; it will never get there, but certain systems can be affirmed as superior to others (either on the basis of predictive accuracy noted above, or even on account of such factors as consistency, cohesiveness, sophistication, the number of phenomena accounted for by each principle etc.).

of course that the instruments employed in observing nature – both natural (our senses) and man-made – were highly fallible. Thus, all the greater was Man's need to compensate for such deficiencies, and to eliminate as many mistakes as possible, whether in terms of measurement, recording of observations, or even the manner in which a particular experiment is interpreted.

In a fashion again reminiscent of his physics, Diderot also emphasised the need for the naturalist to employ creativity, imagination and intuition in the development of hypotheses. Indeed, the method of conjecture briefly sketched out by Diderot in *L'Interprétation de la nature* is demonstrated there with respect to matters of both classical physics and the nature of life. Moreover, we shall see shortly that what may be Diderot's most provocative and profound employment of this method took place while addressing issues related to the latter discipline.

It will have been noted, of course, that the greatest or most obvious continuity of method here is between natural history and physics. It may even be argued that investigation of matters related to life and to living beings was for Diderot nothing other than a branch of experimental physics. An examination of the definition of natural history in Diderot's *Prospectus* to the *Encyclopédie* is very revealing in this capacity. Nominally, 'natural history' is placed under the faculty of 'memory', alongside 'sacred history' and 'civil history'. In this respect, however, natural history pertains primarily to the assemblage of facts about nature and its productions. It is presented essentially as a cataloguing operation.[50] Thus, the discipline of natural history, as it would have been recognised and practised by the Linnaeans, was not even really recognised by Diderot as a science, but rather as an adjunct to or type of history. The attempt to truly understand the 'classic' branches of natural history – botany, zoology, geology etc. – is subsumed for Diderot under the realm of 'physique particulière'. In other words, these pursuits are ultimately seen as facets of the general study of the objects that constitute nature. More significantly, it is in this context of particular physics that the scientist attempts to come to terms with the true nature of these objects. Of everything listed under 'physique particulière' – from the stars, to meteorology, to plants and animals – Diderot said that 'la réflexion a passé à la recherche de leur origine, des causes de leurs phénomènes, etc.'[51] Here, the *Encyclopédie* is true to the traditional structure of Aristotelian physics (albeit in slightly modified, Cartesian form) to which Diderot would have been exposed at university.[52]

The relatedness of Diderot's physical science to his life science extends well beyond matters of definition and methodology however. Much of the

50. *Prospectus*, p.299-302.
51. *Prospectus*, p.308-10.
52. Brockliss, *French higher education*, p.337.

content of Diderot's world, as it is outlined in physico-chemical terms, is entirely relevant to his considerations regarding the nature of life and its history. The list of themes and ideas carried over from other aspects of his study to this one is extensive, but one of its more notable items is Diderot's quest to broaden the accepted definition of matter. As we have seen, Diderot resisted the idea that matter might be conceived simply in terms of extension and motion (or even extension, motion and gravity). Part of his concern here was addressed primarily towards matters of pure physics – that is, considerations about the forces at work in directing the actual movement of particles. Another, closely related part of his concern was relevant to chemistry and the various possible combinations of matter. A further aspect of Diderot's quest was to conceive of matter in such a way that it permitted a materialist explanation of life itself. Thus, an acknowledgement that Man's definition of matter was far too vague and, in being systematised, misconstrued, was put to work for Diderot within what moderns would call the realms of physics, chemistry and biology. All three arguments, all three concerns share the common root of Diderot's desire to revise contemporary understandings of matter itself.

Related to this is Diderot's appreciation of the importance of force and his suspicion, reinforced by the chemistry of Rouelle, that force may well be more important than matter in terms of accounting for all types of phenomena. Likewise, Diderot's hostility towards the ideas of 'essential qualities' and 'final causes' are manifest in his natural history just as they are in his physico-chemistry. In *L'Interprétation de la nature*, for instance, he said: 'Qui sommes-nous pour expliquer les fins de nature? Ne nous apercevrons-nous point que c'est presque toujours aux dépens de sa puissance, que nous préconisons sa sagesse, et que nous ôtons à ses resources plus que nous ne pouvons jamais accorder à ses vues? Cette manière de l'interpréter est mauvais, même en théologie naturelle.'[53]

More than this, there are what may be called four pillars of Diderot's natural philosophical world-view. Each of these have been touched upon in at least one (if not all) of the previous chapters, but only upon consideration here does it become apparent that these ideas are universal aspects of his thought and fundamental to his entire vision of the world; they underlie and condition every aspect of Diderot's natural philosophical outlook.

The first of these pertains to time. This was a contentious issue in mid-eighteenth-century France. Many natural historians, led by Buffon, posited that the age of the Earth was much greater than that stipulated in the Bible. Operating on the presumption that the Earth was originally constituted of molten matter, Buffon heated a sphere of iron and measured the amount of time it took to cool. Extrapolating upon this

53. *Pensées sur l'interprétation de la nature*, p.764.

data to account for the actual size of the Earth, Buffon suggested that it would take 42,964 years and 221 days to be able to touch such a sphere without burning oneself, and 96,670 years and 132 days for it to arrive at 'the point of its [the Earth's] current temperature'.[54] Such an estimation, of course, had the potential to cause tremendous consternation amongst theologians, who insisted that the Earth could be no older than the 6,000 years 'revealed' in the Bible.[55] Although Buffon would later revise his estimate (extending the amount of time, rather than contracting it), he continued to defend the position that the world was much older than had previously been accepted.

Diderot supported Buffon's attribution to nature of a history that was much longer than theological conceptions allowed, and he also seemed to accept Buffon's chronology of the solar system's creation. His acceptance of this chronology, however, was not dogmatic. An example of this is presented in *L'Interprétation de la nature*, in which Diderot stated 'Supposé que la terre ait un noyau solide de verre, ainsi qu'un de nos plus grands philosophes [Buffon] le prétend'.[56] This very general, non-committal espousal of Buffon's hypothesis is indicative of two things. First, it demonstrates that Diderot was not tremendously captivated by geology in and of itself. Specific data related to the physical composition of the Earth, the fossil record or the precise manner in which the solar system was born was of merely casual interest to Diderot.

The second point, which is a veritable extension of the first, is that Diderot was not interested in determining the exact age of the Earth. He was in accord with Buffon's intentions in moving away from the biblical timeline, but was not troubled about the precise number of years determined. This lack of concern reveals that Diderot's perception of time and its role in nature was fundamentally different in its inspiration and perspective from that of Buffon. An excellent example of this is *Le Rêve de D'Alembert*. In the first dialogue, the character Diderot states to D'Alembert that: 'Lorsque le bloc de marbre est réduit en poudre impalpable, je mêle cette poudre à de l'humus ou terre végétale. Je les pétris en ensemble. J'arrose le mélange. Je le laisse putréfier, un an, deux ans, un siècle. Le temps ne me fait rien.'[57]

Thus even Buffon's determinations about the age of the Earth were

54. Roger, *Buffon*, p.409. Although he had addressed the issue of the Earth's history and the creation of the solar system earlier in his career, the results of Buffon's iron ball experiments were not published until 1774, with the appearance of his *Supplément* to the *Histoire naturelle*.

55. More on these issues (albeit as they were played out on the other side of the English Channel) can be found in Charles Coulston Gillispie, *Genesis and geology: a study in the relations of scientific thought, natural theology, and social opinion in Great Britain, 1790-1850* (New York, 1959).

56. *Pensées sur l'interprétation de la nature*, p.735-36.

57. *Le Rêve de D'Alembert*, p.59.

superfluous for Diderot. For him, the history of the world was characterised by inconceivable expanses of time. For all practical purposes it extended indefinitely into both the past and future. Where Buffon and other contemporaries spoke in terms of tens of thousands of years, Diderot spoke in terms of millions.[58] Where Buffon was concerned to determine the precise age of the Earth, Diderot simply wanted to assert that nature has had an inconceivably long period of time to work its productions, and that it will continue to do so in the future.

It may be remarked of course that Diderot's perception of time was not scientifically grounded. Certainly he had no concrete evidence to suggest that the universe may be millions, or billions, of years, or for that matter infinitely old. At best, his presumption that the history of the world extends back into mists that are for all practical purposes beyond human reckoning might be traced to two important considerations. The first is the fossil record, and evidence that forms of life which existed on the Earth in the past were radically different from the forms that exist today. This consideration may then be married to the common-sense observation that while things may change, they do so only at an extremely slow rate. Of course, a number of recent scholars have shown that the role and meaning of fossils in eighteenth-century natural philosophy was ambiguous at best.[59] Taking account of the two premises mentioned above does not therefore lead one to the conclusion that it is necessary to talk in terms of millions of years as opposed to thousands. Ample testimony to this is the fact that the same information was available to Diderot's contemporaries, yet almost all of them choose to interpret the evidence in a different way. It was entirely plausible, at least in eighteenth-century France, for a theologian to argue that fossils were 'planted' by God, or more likely that they were remnants of the biblical flood – perhaps animals that did not make it onto the ark. Likewise, it was possible for the naturalist such as Buffon to agree that the world has taken a long time to acquire its current shape, but that the rate of change that has produced this state of affairs is not nearly as slow as Diderot presumed. It must also be noted that in contemplating the history of the universe Diderot was interested primarily in the development of animal forms, whereas the calculations of Buffon took place in the context of the physical Earth, or geology.

In any case, just as Pascal stared into the infinite, so Diderot stared into eternity. But where Pascal's experience was harrowing, Diderot's perception of time as being virtually unlimited was liberating. It freed his

58. *Pensées sur l'interprétation de la nature*, p.769.

59. See Rhoda Rappaport, 'Geology and orthodoxy: the case of Noah's Flood in eighteenth-century thought', in *The British journal for the history of science* (March 1978), p.1-18. See also Martin Rudwick, *The meaning of fossils: episodes in the history of palaeontology* (New York, 1976).

thought from the bounds of following a specific chronology, and gave him what for all intents and purposes was infinite scope to posit the transformations of nature that he wished. The fact of these transformations was, of course, another important pillar of Diderot's natural philosophy. Positing as he did that change was constant and that nature had a vast expanse of time to carry out its productions, Diderot was led quite naturally to the proposition that tremendous alterations could and did take place in the natural order. While maintenance of this view placed Diderot at odds with many of his contemporaries on yet another front, it was probably not as isolating as his attribution to the universe of potentially unlimited amounts of time. Indeed, Diderot's acknowledgement of the world as dynamic and transformable was but part of a long tradition extending back arguably to the Ancients. Essentially, his claim was nothing more than a subscription to the *natura naturans* school of thought (that nature is a process, and things must be regarded as existing in a state of continuous becoming, rather than being), as opposed to *natura naturata* (which insisted upon the stability and permanence of natural forms). This division had been relatively clear in the Western tradition since at least the Renaissance, and in positing the fundamentality of change Diderot was merely following in the footsteps of Leibniz, Shaftesbury and Maupertuis.[60] The latter two may even be identified as immediate predecessors, as it was probably in translating Shaftesbury's *Essay on merit and virtue* that Diderot first came across an integrated and coherent argument in favour of regarding change as a primary element of nature.[61]

As we have seen, Diderot emphasised the slowness of nature's operations, and of its transformations. He asserted that often change took place so slowly, that is to say by such tiny increments, that it was imperceptible, certainly during the lifetime of an individual human being. In *L'Interprétation de la nature* Diderot points out that: 'La nature est opiniâtre et lente dans ses opérations. S'agit-il d'éloigner, de rapprocher, d'unir, de diviser, d'amollir, de condenser, de durcir, de liquéfier, de dissoudre, d'assimiler, elle s'avance à son but par les degrés les plus insensibles. L'art au contraire se hâte, se fatigue et se relâche. La nature emploie des siècles à préparer grossièrement les métaux.'[62] Yet despite the frequent imperceptibility of change in the world, Diderot was equally staunch about asserting that, given the vast expanses of time available to nature in carrying out its operations, the most profound transformations could and did take place. In *Réfutation sur Helvétius* he

60. Another notable precursor was Fontenelle, who, as Aram Vartanian amply demonstrates, was deeply committed to the idea of the infinite diversity and creativity of a universe in perpetual flux. See Vartanian, *Science and humanism*, p.20, 34 and 39.

61. Dorothy Schlegel, *Shaftesbury and the French deists* (Chapel Hill, NC, 1956), p.43-62.

62. *Pensées sur l'interprétation de la nature*, p.744.

takes issue with the claim that 'La différence de la latitude n'a aucune influence sur les esprits' by arguing 'Je n'en crois rien, ne fût-ce que par la raison que toute cause a son effet, et que toute cause constante, quelque petite qu'elle soit, produit un grand effet avec le temps'.[63]

This theme of change over time, of a universe characterised by perpetual flux is emphasised throughout Diderot's career. In *Lettres sur les aveugles* he has Saunderson invite his interlocutor to imagine a world in which things have not always been as they are now, in which matter moves itself in a veritable cauldron of chaos, and in which the forms presently observed will be succeeded by others.[64] Many years later, in *Eléments de physiologie*, Diderot advances the strong claim that to deny the fundamentality of change in the universe is an absurdity.[65]

There is, however, an important qualification that must be made to Diderot's views on the dynamism of the universe. He did not conceive of change simply as a basic force at work in the world. In this sense his dynamism is very different from that of Leibniz. Where Leibniz posited the existence of monads (metaphysical entities which changed over time in accordance with God's pre-established harmony) for Diderot change was a more physical consideration. Transformations did not take place because they had been willed or because a principle of change actually exists in nature, but simply because change was a fact. Part of the basis for this claim was the observability of change itself. Insofar as dynamism had a theoretical or philosophical basis, however, Diderot's views here were intimately linked to a third pillar of his world-view: the interconnection of all matter in the universe.

Contrary to commonly expressed opinions, such as that of Colm Kiernan, Diderot did not assert that because change is a fundamental fact of reality it occurs in a random or catastrophic fashion.[66] His reference to the accumulation of great effects over time, even at the impetus of the smallest causes, is demonstration enough of this. But an even more potent aspect of his affirmation that the universe is incessantly changing is his vital assertion that everything in nature is ultimately linked. Diderot argued that despite the apparent individuality of everything, all is but part of a whole; the universe is like one great organism.[67]

This point of view is upheld by Diderot's perception of the great chain of being. Throughout his philosophical career, even as late as *Eléments de physiologie*, Diderot asserted that there is a gradation of all living forms, with smooth, almost seamless transitions between each. While he may not

63. *Réfutation sur Helvétius*, p.510.
64. *Lettre sur les aveugles*, p.197-98.
65. *Eléments de physiologie*, Quintili edn, p.137.
66. Kiernan, 'Additional reflections', p.134.
67. 'Animal', p.66.

have affirmed that this interconnectedness was strictly linear (witness Rouelle's demonstration that the three kingdoms of nature are not vertically linked but that the relationship is much more complicated than that), he did accept that everything was connected in some way. This is supported by his proposition that 'rien ne se fait par saut dans la nature'.[68] The principle of interconnection is further emphasised by views expressed in *De la nature*. Not only are there references here to 'la grande chaîne qui lie toutes les choses',[69] but Diderot insists that 'L'indépendance absolue d'un seul fait est incompatible avec l'idée de tout; et sans l'idée de tout, plus de philosophie'.[70]

Ernst Cassirer proposed that for Diderot transformation was the key to seeing nature as a whole.[71] The relationship, however, clearly works the other way as well; for Diderot, the interconnectedness of everything in nature was the reason that it could be affirmed as continually changing. Because, according to Diderot's perception, a 'law' of nature was to be regarded as nothing more than a representation of the 'rapports' which exist between phenomena, it stands to reason that when something changes it will have an impact upon the 'general order'. In other words, Diderot supported the almost tautological proposition that the world is continually changing because the things in it change, even if only with respect to one another. The alteration or transformation of one component has a direct or indirect effect upon everything else. Thus, over the infinitely extended period of time that Diderot allows nature to carry out its works, absolutely everything is bound to be changed eventually. The position is succinctly expressed in *L'Interprétation de la nature*: 'Les phénomènes seraient tous enchaînés, que l'état de chacun d'eux pourrait être sans permanence.'[72] Thus, as Diderot perceived the universe, change did not take place on account of some intelligence or guiding hand, and it did not move in a particular direction towards a goal or final cause. Instead, everything happens on account of causal connections. Because everything is either directly or indirectly connected, a change in one part of the system necessarily impacts upon other parts; as one component of the universe is transformed, so the nature of the whole has been altered. Essentially Diderot saw change taking place simply because it did.

This in turn raises another key feature of Diderot's world-view, namely that the universe is neutral – it does not have to conform to Man's ideas about it. Just as Diderot held that mathematics does not actually exist in nature, so too he maintained that in nature there is no absolute good or

68. *Réfutation sur Helvétius*, p.564.
69. *Pensées sur l'interprétation de la nature*, p.719.
70. *Pensées sur l'interprétation de la nature*, p.722.
71. Cassirer, *Philosophy of the Enlightenment*, p.90.
72. *Pensées sur l'interprétation de la nature*, p.768.

bad, perfection or imperfection. Such judgements, according to Diderot, could only be made with regard to relationships between individuals or sets of individuals, and therefore were entirely relative. It is possible and indeed necessary for the naturalist to speak of success or failure, survival or death. These categories, however, were seen by Diderot as entirely practical and descriptive, separated from inadmissible proclamations such as those about design, intentionality, goal-driven progress or absolute ethical standards. Nature, for Diderot, is blind. In this context, morality is reduced to the role of a 'survival mechanism'; it is nothing more than a set of rules that sacrifices a certain amount of individual liberty in the interests of promoting the cohesiveness, defensibility and therefore survival of a particular group.[73]

This presents another dramatic illustration of how far removed Diderot's conception of nature was from that of many of his contemporaries.[74] The eighteenth century was notable for the pre-eminence of books such as abbé Noël-Antoine Pluche's *Spectacle de la nature* (1732-1759), which sought to demonstrate through the sophistication and elegance of nature's design that it was indeed the production of a benevolent and intelligent being.[75] One of the biggest objectors to this method of course was Buffon, whose ideas about the proper means of regarding nature influenced, or at least vindicated, Diderot's own position here just as they did with regard to the criticism of Réaumur and Linnaeus. As Jacques Roger argues, Buffon felt it essential to proceed as an effective atheist in one's investigation of nature. Although there is dispute regarding the nature of Buffon's personal religious beliefs, and although he never stressed or promoted the cause of atheism for its own sake (being far too comfortable and conservative to upset the establishment in which he himself enjoyed reasonably high standing), it was part of Buffon's natural philosophical methodology to assert that 'good science' is 'secular science'; his view was that in trying to explain phenomena we cannot and should not make appeal to forces existing outside of nature.

But here, as elsewhere, Diderot seems to have gone farther than Buffon,

73. Lester Crocker, 'The Idea of a neutral universe in the French Enlightenment', in *Diderot studies* 21 (1983), p.66.

74. Crocker, in the article indicated above, picks up on the impersonal, godless aspect of Diderot's Nature. As the title of his article suggests however, he asserts that this view was actually part of a burgeoning trend in natural philosophy. He argues that 'many in eighteenth-century France fostered a science that abandoned the pretension to know the meaning of the world' (p.51). It remains, however, that an equally great number of philosophers were still tremendously receptive to and enthusiastic about accounts of the world which afforded God (or even simply ideas about design and final causes) a prominent position.

75. See Benoît de Baere, *Trois introductions à l'abbé Pluche: sa vie, son monde, ses livres* (Geneva, 2001). For more on the natural theological tradition in general (one of the last great exponents of which was Charles Bonnet) see Jean Ehrhard, *L'Idée de nature en France dans la première moitié du XVIIIe siècle*, p.45-59.

or at least to have taken a different tack. His concern was not simply to eliminate considerations of the Supreme Being from all investigations of nature, but, as is suggested above, to stress that nature is consummately unresponsive to all of Man's ideas about it. According to Diderot's natural philosophical view, one of the greatest sins is anthropomorphism. Thus it is that *Pensée 56* of *L'Interprétation de la nature* argues: 'Le physicien dont la profession est d'instruire et non d'édifier, abandonnera donc le *pourquoi*, et ne s'occupera que du *comment*. Le *comment* se tire des êtres; le *pourquoi*, de notre entendement [...] Combien d'idées absurdes, de suppositions fausses, de notions chimériques dans ces hymnes que quelques défenseurs téméraires des causes finales ont osé composer à l'honneur du Créateur?' (*italics* are Diderot's).[76] The only meaning that exists is that which Man creates, and the natural philosopher must always bear in mind that this meaning does not really exist in nature. In attempting to impose some sense of order upon reality, the natural philosopher must be careful not to overvalue the essentiality of his own ideas, and must always try to refrain from reading things into nature, accepting rather that nature simply is, and that human ideas about it are just that – ideas.

Taking into account these pillars of Diderot's natural philosophy, we are now in a position to evaluate some of his more thought-provoking ideas pertaining to the phenomena and nature of life. Traditionally there are two aspects of his thought here that have attracted a great deal of attention: the idea of *sensibilité universelle*, and that of transformism. Between them, these notions have tended to monopolise both interest in and debates about Diderot's natural philosophy, and for good reason. In both cases, Diderot's ideas regarding the matter at hand seem to place him at the cutting edge of eighteenth-century thought. Moreover, *sensibilité* and transformism are the subjects of some of most powerful, dramatic, imaginative and thought-provoking passages to be found in the entire canon of Diderot's works. In being armed with the systematic appraisal of the *philosophe*'s methodology that has been performed over the preceding pages, it is hoped that we will now be in a position to fruitfully revisit both of these crucial ideas. Doing this extends the promise of, at the very least, better comprehending how these notions fit into the framework of his natural philosophy as a whole. At most, such an examination holds out the possibility of revising or refining our perspective on exactly what Diderot was trying to propound with regard to his thoughts on the nature of life.

The idea of *sensibilité*, or sensitivity, was at the centre of much debate in the eighteenth century, particularly during the latter half. At its most general the word denoted for natural philosophers the capacity for feeling or awareness. As such, the idea was distinguished from Albrecht von Haller's 'irritability', which referred only to observed contractibility of

76. *Pensées sur l'interprétation de la nature*, p.765.

muscle fibres. Like Haller's irritability however, sensitivity was commonly seen to be the prerogative of living things. Henri Fouquet's *Encyclopédie* article 'Sensibilité' noted that the property was 'inhérente à la substance du nerf ou des parties nerveuses, a l'exclusion de toutes autres substances du corps'.[77] At the very least then, the capacity for exhibiting sensibility was seen as a basic dividing line between organic and inorganic matter. But what was its true nature? Did the concept of sensibility infer that organic material was fundamentally distinct from the inorganic realm, that it rested on a different set of principles, and that it was therefore subject to different laws? For some, the propensity of living things to 'feel' was not really a propensity at all, but rather an observed effect of more general, physical causes at work within the organism. Others, however, tended to identify this capacity for feeling with a unique principle or series of principles assumed to reside within or underlie organic beings. It was often supposed not just to account for the capacity to feel, but for all of a living being's higher faculties as well, such as thought and consciousness. In this context, *sensibilité* could be conceived as the manifestation of a basic force of nature, a vivifying principle posited to explain the origins and development of life without reference to purely mechanical laws. In this respect, sensitivity is commonly identified with vitalism; the belief that life itself cannot be reduced to anything else but is a fundamental force of nature which is responsible for animating objects.[78]

Diderot's thoughts on the issue were drawn out over a period of at least thirty years. Their development can be traced through a number of his works and letters. His first serious attempt to grapple with the issue was in *L'Interprétation de la nature*, where he deals at length with Maupertuis's discussion of life. Taking up the hypothesis of a certain Dr Baumann of Erlangen (an alias used for Maupertuis), Diderot states that he intends to push the hypothesis as far as it will go. Nominally the purpose of this exercise is to demonstrate Diderot's method of creative hypothesising, to show his readers how the natural philosopher might go about assimilating and assessing certain relevant ideas as part of the project of better grasping the nature of the world. On a more substantive level however, this discussion represents Diderot's foray into the realm of *sensibilité*.

As we have seen, one of the most significant aspects of this discussion is Diderot's rejection of Maupertuis's attribution to matter of the qualities desire, aversion and memory. An explicitly stated reason for casting out this model was that it lead, in Diderot's view, to Spinozism, a stance that

77. Vartanian, *Science and humanism*, p.141.

78. François Duchesneau, *La physiologie des lumières: empirisme, modèles et théories* (The Hague, 1982). Guido Cimino and François Duchesneau, *Vitalisms: from Haller to the cell theory: proceedings of the Zaragoza Symposium: 19th international congress of history of science: 22-29 August 1993* (Florence, 1997). Elizabeth Haigh. *Xavier Bichat and the medical theory of the eighteenth century* (London, 1984). These issues shall be taken up in greater detail in the next chapter.

Diderot was keen to avoid not only on philosophical grounds (due to its inappropriate level of speculation as well as anthropomorphism), but, like his contemporaries, on account of pressure from the establishment as well.[79] More than this, however, Diderot accused Maupertuis's hypothesis of being circular, of merely providing as an explanation a version of what it is trying to explain. Intelligence is accounted for by positing that it simply exists at the most fundamental level, and in a form which does not differ substantially from its macroscopic manifestation. To attribute the qualities of desire, aversion and memory to matter is, Diderot feels, tantamount to hylozoism. In the end, it explains nothing, and worse yet, relies on what must be wholly unsubstantiated speculation.

In other words, Diderot seems to offer what appears to be a very rational, perhaps even traditionally mechanical set of objections to Maupertuis's idea. It is condemned for being baseless, speculative, lacking explanatory power, indulging in anthropomorphism and leading to dangerous conclusions. Significantly however, Diderot did concur with Maupertuis' general goal of broadening the accepted definition of life. He agreed with Dr Baumann's thesis that contemporary mechanical philosophy alone is not sufficient to the task of accounting for life or explaining its operation. Diderot notes that:

> L'auteur [Maupertuis] commence par exposer rapidement les sentiments de ceux qui l'ont précédé, et l'insuffisance de leurs principes pour le développement général des phénomènes [...] l'impossibilité d'expliquer la formation d'une plante ou d'un animal, avec les attractions, l'inertie, la mobilité, l'impénétrabilité, le movement, la matière ou l'éntendue, a conduit le philosophe Baumann à supposer encore d'autres propriétés dans la nature [...] il a pensé que tous ces systèmes peu philosophiques n'auraient point eu lieu, sans crainte mal fondée d'attribuer des modifications très connues à un être dont l'essence nous étant inconnue, peut être par cette raison même et malgré notre préjugé, très compatible avec ces modifications.[80]

Thus, Diderot was in accordance with Maupertuis regarding the general goal, even if he firmly disagreed regarding the nature and terms of the necessary modifications to Man's conception of matter.

Diderot's relationship to the views of many other contemporaries was similar. Buffon also attempted to modify conceptions of matter by positing a 'moule intérieure' or 'interior mold' which was responsible both for investing certain material with life and determining its form.[81] Diderot objected to this notion on the grounds that it introduced a fundamental break in the great chain which links everything together. By

79. A Spinozist view being intimately associated with atheism. Not until the arrival of nineteenth-century Romanticism was Spinoza's confluence of God, Nature, Substance and the Universe truly appreciated in any kind of positive way. See Israel, *Radical Enlightenment*.

80. *Pensées sur l'interprétation de la nature*, p.756.

81. Hall, *Ideas of life and matter*, vol.2, p.12.

positing an unbridgeable gap between 'living' and 'dead' matter, Buffon violated Diderot's fundamental principle of unity. Adamant that there should exist no such rifts in nature, Diderot was unwilling to accept Buffon's vision regarding the basis of life.

Diderot's own solution to these problems is first suggested in *L'Interprétation de la nature*. In *Pensée 51*, the fragment which succeeds his critique of Maupertuis, Diderot contemplated a potential alternative to the vision of Dr Baumann. He suggested that Maupertuis's hypothesis would be more defensible if the notion of molecules possessing desire, aversion and memory was dropped and replaced with the simpler idea that they are infused with 'muted sensibility'. In a short but tantalising passage Diderot remarked:

> Il fallait se contenter d'y supposer une sensibilité mille fois moindre que celle que le Tout-Puissant a accordée aux animaux les plus stupides et les plus voisins de la matière morte. En conséquence de cette sensibilité sourde, et de la différence des configurations, il n'y aurait en pour une molécule organique quelconque qu'une situation la plus commode de toutes, qu'elle aurait sans cesse cherchée par une inquiétude automate, comme il arrive aux animaux de s'agiter dans le sommeil, lorsque l'usage de presque toutes leurs facultés est suspendu, jusqu'à ce qu'ils aient trouvé la disposition la plus convenable au repos. Ce seul principe eût satisfait d'une manière assez simple et sans aucune conséquence dangereuse aux phénomènes qu'il se proposait d'expliquer.[82]

Despite the lack of elaboration, this contemplation represents the true genesis of Diderot's attempts to grapple with the issue of life. It is the beginning of his own formulation of the concept of *sensibilité universelle*. Several years later, in a letter to Charles Duclos (1704-1772), the notion is expressed in a more advanced and confident fashion: 'selon moi, la sensibilité est une propriété universelle de la matière; propriété inerte dans les corps bruts, comme le mouvement dans les corps pesants arrêtés par un obstacle; propriété rendue active dans les mêmes corps par leur assimilation avec une substance animale vivante'.[83]

It was not, however, until 1769 that Diderot's idea achieved its most complete formulation. In *Le Rêve de D'Alembert* he outlined what is one of the eighteenth century's most astonishing, thorough and original theories regarding the nature of life. The work posits that *sensibilité* is a universal property of matter – part of the very definition of matter itself. Diderot averted Maupertuis's hylozoism by arguing that this quality is not equally present in all matter, but that it comes in degrees, such that in a piece of seemingly inert material (like a stone), the amount of *sensibilité* is so small as to be almost negligible, whereas in living matter, it is present to a very high degree. In other words, the categories of animate and inanimate matter are never absolute; 'dead' matter (or what a Cartesian

82. *Pensées sur l'interprétation de la nature*, p.760.

83. 'Lettre à Duclos' (10 October 1765), *OC*, vol.5, p.951.

would have referred to as dead matter) does in fact possess a trace amount of the quality or property that invests 'living' matter with life (even if the true nature of this quality cannot be known, and can only by referenced in the vaguest of ways). *Sensibilité* therefore, according to Diderot's conception, is like potential energy – it is latent in all matter. Moreover, just like potential energy, it always has, in theory at least, the capacity to be realised and become active. This activation of latent *sensibilité* is, the *Rêve* tells us, simply a matter of removing barriers. To illustrate this point Diderot used the example of nutrition: 'Oui, car en mangeant, que faites-vous? Vous levez les obstacles qui s'opposaient à la sensibilité active de l'aliment; vous l'assimilez avec vous-même; vous en faites de la chair; vous l'animalisez; vous le rendez sensible; et ce que vous exécutez sur un aliment, je l'exécueterai, quand il me plaira, sur le marbre.'[84]

Of course, Diderot still maintained that the organisation of matter had a great deal to do with life as well.[85] In this respect, his view rather echoes that expressed by La Mettrie in *L'Homme machine*. While sharing a monist materialist world-view with the infamous doctor however, Diderot did not seem to place as much emphasis as La Mettrie in terms of attributing life to a principle of organisation. Another, perhaps better, way of putting it would be to say that La Mettrie was more reluctant than Diderot to be drawn with regard to discussing the possibility or nature of any force(s) that may reside within matter. Although speaking occasionally of the animation of matter, or of its spiritualisation, La Mettrie comes across as more agnostic on the issue than probably any of his contemporaries. He was not interested in spiritualising matter (as one might say of many Montpellier vitalists, or Maupertuis, or even Diderot, to an extent), but in materialising the soul. Given this, the abiding impression created by *L'Homme machine* is of a distinct emphasis simply on the importance of organisation. It is in this connection, of course, that the work proposes regarding human beings as machines. In even just articulating a notion of *sensibilité*, however, Diderot goes at least a little bit further than La Mettrie in terms of the specificity of his conception. According to Diderot, life and consciousness were dependant upon something immanent in matter, a vital principle or force the ultimate cause of which could not be known. In other words, Diderot was at pains to argue that pieces of matter, appropriately organised, would not in themselves bring about life, but that this relied also on the presence and appropriate level of *sensibilité*.

84. *Le Rêve de D'Alembert*, p.58.

85. This presumption may be connected to Rouelle's insistence that while the material of the 'règne minéral' can be created simply by throwing together the appropriate ingredients in proper proportion, the material of both the 'règne végétal' and 'règne animal' must be combined not only in correct proportion, but in a certain configuration as well (see chapter 3).

Sensibilité, like motion, was not simply accidental to matter as something that can arise given the proper organisation, but was part of its very definition.[86]

Thus, in *Le Rêve de D'Alembert,* Diderot sketched for himself a theory of life distinguished from all of his peers. He was able to distance himself from the hylozoism of Maupertuis, La Mettrie's over-emphasis on organisation (and the corresponding portrayal of man as, quite literally, a machine), as well as Buffon's explicit separation of the organic and inorganic realms. If anything, his thought might appear closest to that of certain Montpellier physicians, in particular Théophile de Bordeu (1722-1776).[87] This debt is most clearly revealed, of course, by Diderot's insertion of Bordeu as a character in the dialogues which comprise *Le Rêve de D'Alembert.*

Diderot borrowed both analogies and concepts from Bordeu. One of the most prominent examples is what has been described as the 'dual-purpose' of Diderot's concept of *sensibilité.* In one sense, the word refers simply to the vitality which is inherent – to at least some degree – in all matter. In another sense, however, *sensibilité* refers to the operations exhibited by that property in its higher or more active states. Responsiveness, growth, nutrition and reproduction and for certain beings consciousness are all expressions of *sensibilité*, they are all attributes of beings that may be described as 'sensible'. In other words, Diderot also used *sensibilité* as a description of the set of features manifested by organised forms of life. It refers to the characteristics belonging to and indelibly defining the status of a living being. To say that an organism is 'sensible' is to say that it is alive, and this attribution to a being of such unifying *sensibilité* cannot be reduced in its essence. To borrow an example that Diderot used in the *Rêve*, it is analogous to saying that we can cut a circle in half, but we cannot have half of roundness. The whole is greater than the sum of its parts.[88] *Sensibilité* can refer then, in Diderot's catalogue of natural philosophical terms, to both a primitive quality inherent in matter and the outward manifestation of that quality.

To modern commentators, the main problem with Diderot's idea is

86. These issues will be addressed more fully in the next chapter, which deals with theories surrounding the nature and operation of the human body (See Ch. 5, p. 196-209). For more on La Mettrie see Claude Morilhat, *La Mettrie; un materialisme radical* (Paris, 1997). Kathleen Anne Wellmann, *La Mettrie: medicine, philosophy, and enlightenment* (Durham, NC, 1992). and Aram Vartanian's critical edition, *L'Homme machine: a study in the origins of an idea* (Princeton, NJ, 1960).

87. For more on Bordeu, see chapter 5, p.178-82.

88. Of course, proclaiming that the whole is greater than the sum of its parts does not necessarily mean that it has the capacity to direct the activity of those parts, or even that the operation of the whole cannot be explained in terms of the relations between the parts. See Timo Kaitaro on 'holistic wholes' versus 'meritistic wholes' in *Diderot's holism: philosophical anti-reductionism and its medical background* (Frankfurt, 1997), p.25. This significance of this distinction will manifest itself even more in the next chapter.

that it seems consummately unscientific. There is no empirical evidence for the notion of *sensibilité*; instead the impetus to Diderot's theory appears to be the inadequacy of mechanical systems to account for life (and as we have seen, even Diderot wondered on occasion whether or not there may come a day when such systems could achieve greater explanatory success). The *Rêve*, therefore, has been frequently passed off as a work of bold, imaginative, occasionally perceptive speculation. But to truly grasp the significance of the work, we must view it firmly within the context of not only Diderot's age, but that of his own philosophy of science. Upon doing this, *Le Rêve de D'Alembert* takes on a very different character.

Aram Vartanian, more than any other recent commentator, has grappled with this aspect of *Le Réve de D'Alembert*. Enriching his argument with a discussion of the semantics surrounding the verb 'rêver', Vartanian convincingly demonstrates that the *Rêve* was a work intended by its author to be read on a variety of levels. Perfectly in tune with his larger thesis about the synthesis between science and humanism in eighteenth-century French thought, Vartanian proclaims that *Le Rêve de D'Alembert* is at once a work of poetry and science, employing the latitude of the former to develop ideas which remained unsubstantiated in terms of the latter.[89] While this interpretation is meritorious, we need not suppose that the *Rêve* was intended by Diderot as a work of poetry used to explore the ramifications of a certain scientific theme.[90] It can just as easily be regarded as a veritable scientific hypothesis intuited by a genius philosopher and as yet awaiting firmer empirical confirmation. Indeed, it is one of the most perfect examples of Diderot's method of creative conjecture as outlined sixteen years earlier in *L'Interprétation de la nature*.

Thus, there does exist the possibility that Diderot's hypothesis of *sensibilité universelle* was intended to incorporate a certain amount of truth, or at least points the way towards a more accurate explanation of the basis and nature of life. It also provides the impetus to and guidance for further research; it operates as something that, if and when employed, might yield useful results. His view is both highly speculative and yet very powerful. That Diderot accepted the ideas contained within the *Rêve* as such is revealed by a variety of sources. In a letter to Sophie Volland, for instance, he said of *Le Rêve de D'Alembert*: 'Cela est de la plus haute extravagance et tout à la fois de la philosophie la plus profonde. Il y a quelque adresse à avoir mis mes idées dans la bouche d'un homme qui rêve.'[91]

Diderot's *Réfutation d'Helvétius* lends further weight to this perception of the theory of *sensibilité universelle* as a useful hypothesis. On several

89. Vartanian, *Science and humanism*, p.116-24.

90. Although by regarding it in this sense, Diderot may be seen as a precursor of Erasmus Darwin (1731-1802). See Maureen McNeil, *Under the banner of science: Erasmus Darwin and his age* (Manchester, 1987).

91. 'Lettre à Sophie Volland' (31 August 1769), *OC*, vol.13, p.895.

occasions, Diderot addresses the idea with extreme wariness. He states, for instance: 'Il faut que les notions de matière, d'organisation, de mouvement, de chaleur, de chair, de sensibilité et de vie soient encore bien incomplètes.'[92] Shortly thereafter, and even more tellingly, Diderot remarks that 'la sensibilité générale des molécules de la matière n'est qu'une supposition, qui tire toute sa force des difficultés dont elle débarrasse'.[93] Such circumspection also appears in *Réfutation d'Hemsterhuis* and in *Eléments de physiologie*. It is tempting to proclaim that these works are more strictly 'scientific' than the *Rêve*, since they are more inclined to stay within the bounds of that which is observable and demonstrable. Nevertheless, these works operate in conjunction with the *Rêve* to illustrate essential aspects of Diderot's natural philosophical method. *Le Rêve de D'Alembert* represents the indispensable process of developing hypotheses, and *Réfutation d'Helvétius*, *Réfutation d'Hemsterhuis* and *Eléments de physiologie* represent the equally necessary process of evaluating these hypotheses from a sober, 'scientific' perspective, attempting to ascertain their place in Man's body of knowledge and integrate them with what may already be affirmed.

Taken together, these four works exemplify Diderot's natural philosophy at its best – bold and innovative, and yet appropriately cautious. As an instance of attempting to integrate a particular hypothesis into a larger theory of life, Diderot's work and thought on *sensibilité* can be seen as an attempt to point the way forward, to 'push things as far as they will go' while remaining eminently self-conscious and aware of self-imposed rules of method. *Sensibilité universelle* was, to Diderot, simply the most philosophically defensible explanation for the nature of life and the means by which it operates. It was not an absolute creed, but the best possible hypothesis given the available evidence. Diderot acknowledged that this hypothesis left much room for error and would undoubtedly require substantial revision, but it was nonetheless a start.

We might see in very much the same light that which may be the most remarkable aspect of Diderot's 'natural historical' thought: his 'proto-evolutionary' ideas. Like the idea of *sensibilité*, this is an element of Diderot's work that has drawn no small amount of attention over the years, yet profound discrepancies in the relevant historical literature indicate that it still has the capacity to perplex and to ignite debate. Lester Crocker, for example, argues that while Diderot's 'proto-evolutionary' thought is severely lacking in several important respects, it is both novel and at the forefront of eighteenth-century French speculation on the issue.[94] Pierre Naville, comparing Diderot's transformism to that of

92. *Réfutation d'Helvétius*, p.492.

93. *Refutation d'Helvétius*, p.492.

94. Lester Crocker, 'Diderot and eighteenth-century French transformism', in *Forerunners of Darwin, 1745-1859*, ed. B. Glass, O. Temkin and W. Straus (Baltimore, MD, 1959), p.115-42.

his contemporaries, calls it 'brilliantly developed'.[95] Colm Kiernan, on the other hand, asserts that the theory of evolution derives more from a fixist perception of species than from the transformism associated with Diderot,[96] while much earlier in the twentieth century Ferdinand Paitre maintained that the true ancestor and greatest exponent of transformism in France was not Diderot but Robinet, whose own *De la nature* (1766) developed in strikingly bold fashion themes that may be described as 'proto-evolutionary'.[97]

Diderot was of course inspired by others in the development of his thought in this area, but as usual drew upon a broad variety of sources and interwove these with a range of his own core ideas to articulate a remarkable thesis. As early as *Lettre sur les aveugles* Diderot called upon the themes of the dynamism, interconnectedness and neutrality of the universe to propose the probability that species change form over the course of many successive generations. It is precisely this possibility with which Diderot has Saunderson challenge his interlocutor's vision of a created, fixed and benevolent universe:

> Imaginez donc, si vous voulez, que l'ordre qui vous frappe a toujours subsisté; mais laissez-moi croire qu'il n'en est rien; et que, si nous remontions à la naissance des choses et des temps, et que nous sentissions la matière se mouvoir et le chaos se débrouiller, nous rencontrerions une multitude d'être informes pour quelques êtres bien organisés [...] Je puis vous demander, par exemple, qui vous a dit à vous, à Leibnitz, à Clarke et à Newton, que dans les premiers instants de la formation des animaux, les uns n'étaient pas sans tête et les autres sans pieds? [...] que les monstres se sont anéantis successivement; que toutes les combinaisons vicieuses de la matière ont disparu, et qu'il n'est resté que celles où le mécanisme n'impliquait aucune contradiction importante et qui pouvaient subsister par elles-mêmes et se perpétuer.[98]

This vision is reminiscent of certain ancient perspectives on the universe, and some would even argue that it draws its primary inspiration from Lucretius.[99]

The idea of species developing over time is propelled further in *L'Interprétation de la nature*. Here, Diderot not only contemplated the likelihood that species are either irrevocably changed or extinguished over the course of time, but gave serious consideration to the hypothesis that

95. Pierre Naville, *D'Holbach et la philosophie scientifique au XVIIIe siècle* (Paris, 1967), p.262.

96. Kiernan, 'Additional reflections', p.138.

97. Robinet's work incorporated ideas about living matter, nature as a collection of individuals, and the progression of forms from the relatively simply to the relatively complex, with the composition of the 'ideal Man' being the apparent goal of this general tendency. See Fernand Paitre, *Diderot biologiste*, p.92. See also Crocker, 'Diderot and eighteenth-century French transformism', p.134.

98. *Lettre sur les aveugles*, p.197-98.

99. Roger, *Life sciences in eighteenth-century French thought*, p.520.

all animals are ultimately derived from a single being, or prototype. Interestingly, just as with Diderot's initiation into matters of *sensibilité*, the natural philosopher whose ideas are at the centre of discussion here is none other than Maupertuis. Contrary to Buffon, Maupertuis believed in the possibility, if not the probability, of such a prototype animal having existed. Diderot was rather circumspect and non-committal about the issue, but found it a very intriguing if not reasonable hypothesis. In *Pensée 12* he outlined then proceeded to comment upon Maupertuis' idea:

> Quand on considère le règne animal, et qu'on s'aperçoit que parmi les quadrupèdes, il n'y en a pas un qui n'ait les fonctions et les parties, surtout intérieures, entièrement sembables à un autre quadrupède, ne croiait-on pas volontiers qu'il n'y a jamais eu qu'un premier animal prototype de tous les animaux dont la nature n'a fait qu'allonger, raccourcir, transformer, multiplier, oblitérer certains organes? [...] Mais que cette conjecture philosophique sont admise avec le docteur Baumann comme vraie, ou rejetée avec M. de Buffon comme fausse, on ne niera pas qu'il ne faille l'embrasser comme une hypothèse essentielle au progrès de la physique expérimentale, à celui de la philosophie rationnelle, à la découverte et à l'explication des phénomènes qui dépendent de l'organisation.[100]

Diderot was quite willing to admit Maupertuis's proposition, even if only as a working hypothesis. Elsewhere in *L'Interprétation de la nature* he adopts a very similar attitude with respect to the prospect of extinction for entire species, remarking that 'De même que dans les règnes animal et végétal, un individu commence, pour ainsi dire, s'accroît, dure, dépérit et passe; n'en serait-il pas de même des espèces entières?'[101]

Of course, this 'proto-evolutionary' view is quite negative, in the sense that it does not proclaim evolutionary development as a progressive force at work in the world, but merely sees no reason why it should not be the case that some species are radically transformed and indeed many wind up dying out completely. This might create the sense that Diderot is in fact more of a philosopher here than a natural philosopher per se. Such a sense is reinforced by Lester Crocker's article, which asserts that Diderot's thoughts on the issue were lacking at this stage in a number of crucial respects. The first, is that Diderot accepted spontaneous generation, an espousal which, in the eyes of many scholars, is tantamount to embracing a catastrophic (and therefore unscientific) world-view; that is, a view that perceives nature as operating not according to slow, steady, predictable and law-governed transformations, but by huge and random leaps.[102] Second, Crocker criticises Diderot's failure to capitalise on Maupertuis's thoughts about heredity, which would have lent greater weight to any

100. *Pensées sur l'interprétation de la nature*, p.723.

101. *Pensées sur l'interprétation de la nature*, p.769.

102. For a perfect example of this equation of spontaneous generation with 'catastrophism' (in the most philosophically derogatory sense), see Kiernan, 'Additional reflections'.

theory of how and why species transform.[103] Finally, he notes Diderot's inability to elucidate a mechanism by which evolution might work (such as Darwinian natural selection).[104]

A close investigation of all Diderot's 'proto-evolutionary' statements however (particularly as they were to develop throughout his career) presents a picture more favourable to him as a serious and systematic natural philosopher. First, spontaneous generation can be invoked merely to account for the initial origins of life. In a universe bereft of God (such as Diderot's) there must be some presumption that in the beginning life arose from non-life. As Jean-Baptiste Lamarck (1744-1829) declared: 'Once the difficult step of admitting spontaneous generation is made, no important obstacle stands in the way of being able to recognise the order and origins of the different productions of nature.'[105] This does not imply of course that things have to spring into existence on a regular basis, or that generation is a random process. Indeed, despite the fact that Diderot seems never to have firmly decided upon a preferred system of generation,[106] it is clear that he believed that generation followed some manner of regular course or 'rule of law'. That this was the case is demonstrated in *Pensée 32* of *L'Interprétation de la nature* in which Diderot asserted not only that both sexes co-operate in the process of creating life, but 'si le môle est le résultat d'une combinaison telle que je la suppose, cette combinaison aura ses lois aussi invariables que celles de la génération'.[107]

As this implies, Maupertuis' notion of heredity, or some concept of heredity does come into play here. Of course, the biological foundation for Maupertuis's ideas were quite different from our own, and therefore a few words about the concept and functioning of heredity according to Maupertuis is appropriate here. The idea was originally developed as part of a response to a problem that prevailing eighteenth-century theories of reproduction and development (particularly preformationism) had been unable to resolve: that of albinism amongst negroes. In a letter entitled *Nègre-Blanc* (1744) Maupertuis commenced his speculations and, after an exhaustive study of polydactyly (extra digits), concluded that such characteristics were in fact inherited. Part of the basis for his explanation was that children were formed not simply on account of the mixing of fluids from both parents, but that these fluids actually contained different sorts of particles, each destined to be responsible for

103. Crocker, 'Diderot and eighteenth-century French transformism', p.131. Jean Mayer also lays this criticism against Diderot's evolutionary theory (Mayer, *Diderot, homme de science*, p.258).

104. See Crocker, 'Diderot and eighteenth-century French transformism', p.131.

105. Richard Burkhardt, 'The Inspiration of Lamarck's belief in evolution', *Journal of the history of biology*, 5 (1972), p.434. Burkhardt has also authored *The Spirit of system: Lamarck and evolutionary biology* (Cambridge, MA, 1977).

106. See *Eléments de physiologie*, Quintili edn, p.242-55.

107. *Pensées sur l'interprétation de la nature*, p.734.

forming a different part of the body. The proper union of these particles then became for Maupertuis the basis of inheritance. Intriguingly, his experiments also suggested to him that occasionally 'ancestral particles' (those of a grandparent, for example) would dominate over those of the immediate parents – making hereditary characteristics something traceable over generations. Most strikingly however, Maupertuis suggested, in very mechanistic fashion, that accidents occurring during the process by which the particles contributed by mother and father are mixed lead to the improper development of the foetus and the birth of a 'monster'.[108]

As early as *L'Interprétation de la nature*, Diderot was at least vaguely cognisant of the necessary resemblance between parents and offspring.[109] Later, in *Le Neveu de Rameau* he makes frequent reference to heredity in statements such as 'a fibre granted to me', 'something hereditary in it', 'fathers' blood and uncles' are the same' and the 'paternal molecule'.[110] In *Eléments de physiologie* the role of the mother is observed with regard to the dispositions, tastes and organic aptitudes that she passes along to her child.[111] *Le Rêve de D'Alembert* also provides examples of a belief in heredity, with its mention of hereditary defects, predominant versus submissive bequeathment (to the constitution of the network or 'réseau' that represents the initial state of a living being), and the observation that certain characteristics can skip one generation only to show up again in the next.[112]

Clearly then Diderot had been aware of the notion of heredity, and incorporated within his thought the idea that characteristics are passed on from parents to children in some sort of regular (though unknown) way, and that in this fashion a certain continuity of species is preserved. To be sure, without the concept of heredity, the entire idea of species would be impossible. While Diderot often argued that all things are ultimately individual, and that the classification of beings is an artificial act, he did acknowledge that there were species. Again, *L'Interprétation de la nature* provides the initial articulation, as *Pensée 50* makes reference to 'la conservation des espèces' in conjunction with 'la ressemblance [of chil-

108. Hall, *Ideas of life and matter*, vol.2, p.20-24. Maupertuis's evidence for and delineation of ideas about the manner in which inheritance operates is distributed throughout several of his works: *Venus physique* (1745), *System de la nature* (1751) and a letter entitled 'Sur la generation des animaux' (1752). See Pierre-Louis Moreau de Maupertuis, *Œuvres de Maupertuis*, ed. Jean-Marie Bruyset (Lyon, Jean-Marie Bruyset Imprimeur-Libraire, 1758).

109. This is manifested both in the above comments regarding the co-operation of parents in the production of children (*Pensée 32*) and in *Pensée 50*, where he notes that species are conserved on account of the resemblance of parents to children.

110. *Le Neveu de Rameau*, in Jonathan Kemp, *Diderot; interpreter of nature* (New York, 1943), p.310-11.

111. *Eléments de physiologie*, Quintili edn, p.255-64.

112. *Le Rêve de D'Alembert*, p.110-13. There is also reference here to 'bundles of threads', which the more fanciful historian of science might acknowledge as a striking (if lucky!) foreshadowing of chromosomes.

dren] des parents'.[113] More telling perhaps, is a letter of 2 September 1762 from Diderot to Sophie Volland in which the former proclaims that 'nature has formed Man to perform two functions: conservation of the self, and the propagation of the species'.[114] Following the basic definition of Buffon, species is understood by Diderot as a group of beings which is capable of reproducing amongst itself, and thereby creating further (fertile) beings which, while not the same, will be very similar.[115]

The key for Diderot, however, was that while definable to a certain degree, species can and do change; by a long, very slow process of transformation, they are inevitably modified. Crucially Diderot suggested the means by which this modification takes place. In an argument parallel to the suggestion that the form of a being is determined by both internal and external causes, Diderot posited that species are changed by both their environment (or in response to their environment) and by some internal impetus or accident. With regard to the former case, Diderot asserted that species are affected by factors such as climate, diet, latitude etc. The premier example of this is *Réfutation d'Helvétuis*, which, after all, is a defence of the thesis that education is not the sole determining factor in making human beings what they are, but that nature has just as big a role (if not a bigger role) to play. While Diderot acknowledged that assertions regarding the precise effects of climate are (as yet) too vague to say much, he nevertheless sought to emphasise that it did have some impact on human development. In *Réfutation d'Helvétius* he wonders:

> Croit-on qu'il soit indifférent aux habitants d'une contrée, à leur manière de se nourrir, de se vêtir, de s'occuper, de sentir, de penser, d'être humide ou sèche, couverte de forêts, ou découverte, aride et montagneuse, plate ou marécageuse, plongée dans des nuits de dix-huit heures, et ensevelie sous les neiges pendant huit mois? [...] Si le climat et les aliments influent sur les corps, ils influent nécessairement sur les esprits.[116]

Given this conjunction between habit, environment and human development, it is unsurprising that *Voyage en Hollande* begins with a geographical description of the country, noting the manner in which physical setting and climate impact upon the inhabitants. The first paragraph of the first chapter makes the following observation about the larger forces which guide the destiny of the animal known as Man: 'On dirait que les peuples, soumis comme les autres corps à l'action de la force centrifuge, sont constamment entraînés des pôles vers l'équateur où ils iraient se presser sur une même zone, s'ils n'en étaient écartés par mille causes diverse, et s'ils ne devenaient de plus en plus stationnaires à mesure qu'ils

113. *Pensées sur l'interprétation de la nature*, p.757.
114. *OC*, vol.5, p.734.
115. See Roger, *Buffon*.
116. *Réfutation d'Helvétius*, p.510-11.

descendent.'[117] This view is reinforced in works such as *Eléments de physiologie*, which argues in favour of environmental determinism, and posits a firm causal relationship between factors such as climate and its effect on people.

Further insight into Diderot's ideas about the manner in which species develop over time is to be found in his suggestion that there exists a veritable interdependence between organs and needs. He asserted that the physical needs of an animal dictate, to some degree, the development of its organs, which in turn changes its precise needs. In other words, the physiological structure of an animal limits its functions and determines its needs, but this structure will also adapt to circumstances in the interests of survival. This idea is suggested most explicitly in the *Rêve*, though perhaps most thoroughly in *Eléments de physiologie*, where it is proposed that organs develop according to needs (which they, presumably, have a role in stipulating).[118] The concept of 'need' is defined very broadly here. In one sense, need can be construed as the requirements of an animal as dictated by its physical form. All animals, for example, require nourishment, but some require much more food than others, and they require food of a different type. In another sense, however, Diderot's statement that diverse needs lead to the production of diverse animals seems to suggest that need can also be interpreted as 'that which circumstances provide'.[119] To borrow a Lamarckian example, it might be said that a giraffe's need for food in an environment in which there is a great deal of competition for sustenance at 'ground level' led to the production of its extremely long neck. Despite the ambiguities of this situation, the most important point here is that Diderot certainly felt that the influence of needs on an animal's physiology could produce organs or at the very least transform them.[120] There existed then, for Diderot, tremendous scope by which a species could change in response to its environment. Of course, the argument here is not Darwinian, but very close to Lamarck's theory regarding the inheritance of acquired characteristics.[121]

Nevertheless, there is more to Diderot's proto-evolutionary thought than this, for he did also insist that form is determined and occasionally change imparted on an internal basis. Despite acknowledgement that general rules of generation must exist, Diderot was, like Maupertuis, at pains to demonstrate that things often go wrong. In his universe, no two beings are exactly alike; every individual of every species is at least a little

117. *Voyage en Hollande*, p.335.

118. *Eléments de physiologie*, Quintili edn, p.137-38.

119. *Eléments de physiologie*, Quintili edn, p.137-38.

120. *Eléments de physiologie*, Quintili edn, p.138.

121. Lamarck's evolutionary theory, as propounded in his *Philosophie zoologique* (1809), maintained that animals acquire certain characteristics during their lifetime which are then passed on to their offspring.

bit different in terms of their composition.[122] Thus, the physical structure of people is modified, even if only by 'insensible degrees', and these modifications occur as a result of veritable accidents in the process of generation. That nature is responsible for creating these alterations in animals from the moment of conception, and not simply through the medium of environmental influence, is witnessed in *Le Rêve de D'Alembert*'s aforementioned acknowledgement of 'hereditary defects'.[123] These modifications can, over the course of huge intervals of time, result in massive changes to a species.[124] It is in this way, presumably, that some animals occasionally wind up bearing superfluous organs.[125] The internal impetus for such variation is even suggested as early as *L'Interprétation de la nature*, where it is argued that 'Il semble que la nature se soit plu à varier le même méchanisme d'une infinité de manières différentes. Elle n'abandonne un genre de productions qu'après en avoir multiplié les individus sous toutes les faces possibles.'[126] This view is tantamount of course to mutation, although where the modern scientist would speak in terms of genetic faults, Diderot spoke in terms of 'nature's diverse modifications'.[127]

This touches upon Diderot's profound interest in monstrosity.[128] In a world such as his, where no two beings are exactly alike, everything is 'more or less sick', everything is a 'monster'. Of course, in Diderot's neutral universe, monstrosity is not necessarily a bad thing; how could it be, if there is truly no such thing as absolute 'normality'? Diderot maintained that individuals, monsters and species could not be judged in a straightforward or absolute sense, he perceived all of these things as relative to one another. The only criterion that the naturalist can apply in evaluating beings or groups of beings is success or failure, that is, their ability to survive. If one 'monster' or species outlasts another 'monster' or species then it can be said to have been more successful; not 'better' or 'more perfect', but merely more successful. As Diderot stated in *Eléments de physiologie*, 's'amender, se détériorer sont des termes relatifs aux individus d'une espèce entre eux et aux différentes espèces entre elles'.[129] It is in this sense, as Lester Crocker has observed, that for Diderot morals are reduced to being a 'survival mechanism'.

122. 'Animal', p.72.

123. *Rêve de D'Alembert*, p.110.

124. *Rêve de D'Alembert*, p.91.

125. *Eléments de physiologie*, Quintili edn, p.266.

126. *Pensées sur l'interprétation de la nature*, p.722.

127. *Réfutation de Hemsterhuis*, p.57.

128. For more on early modern ideas about monstrosity see Lorraine Daston, *Wonders and the order of nature, 1150-1750* (New York, 1998). Regarding Diderot specifically, see Emita Hill, 'Materialism and monsters in *Le Rêve de D'Alembert*'; in *Diderot studies* 10 (1968), p.67-93; J.-L. Fischer, 'L'Encyclopédie présente-t-elle une pré-science des monstres?', in *Recherche sur Diderot et l'Encyclopédie*, 16 (1994), p.133-52.

129. *Eléments de physiologie*, Quintili edn, p.265.

This is the pinnacle of Diderot's proto-evolutionary thought: that the only thing which truly matters is the ability of a being (or species) to exist within the 'general order', that is to say, survive alongside everything else that exists at that moment in history. The 'relatively normal' is nothing other than the 'relatively successful', that which survives and is able to pass on its characteristics. Again we might look at Diderot's letter to Sophie, which notes that the two primary functions of an organism are conservation of self and the propagation of species.[130] The 'relatively abnormal' or 'monstrous' is simply that which is so weak, unviable or unsuited to existence within the current scheme of things that it perishes. The strong survive, and the weak or ill-adapted die. This point is one that Diderot sought to emphasise in almost all of his works which have anything to do with the grand scheme of nature. *Les Aveugles*, *L'Interprétation de la nature*, *Réfutation de Hemsterhuis*, *Le Rêve de D'Alembert* and *Eléments de physiologie* all drive this point home forcefully and unremittingly. The last of these works questions why we should not regard all beings as monsters (more or less durable), and why nature should not, over a long period of time, exterminate entire species. Diderot maintains here that everything in the universe is, in a sense, monstrous, and that those beings which we dub monstrous are simply the ones that diverge more significantly from the rest and that (usually) are incompatible with the 'general order'. Since, as we have seen, the general order is continuously shifting, absolutely nothing is safe or immutable. Ultimately all organisms will be transformed into something different from what they are now, or they will perish.[131]

Thus, all of the seeds of a full-blown theory of evolution seem to be contained within Diderot's thought. He dealt with notions of heredity and mutation, and proposed the idea that successful mutations or adaptations survive to pass on their characteristics while the unsuccessful ones simply die out. Crocker and Mayer both argue however that Diderot's proto-evolution is still non-progressive, in that it lacks what Crocker calls the 'finalistic optimism' of Darwin's theory.[132] To a large extent this is true, as Diderot suggests that the sword can cut both ways; things can devolve just as easily as they can evolve. *Le Rêve de D'Alembert*, for example, wonders whether or not species may degenerate back towards an inert mass of particles, and if the mighty elephant may one day regress

130. *OC*, vol.5, p.734.

131. *Eléments de physiologie*, Quintili edn, p.265.

132. Crocker, 'Diderot and eighteenth-century French transformism', and Mayer, *Diderot, homme de science*, p.228. The degree to which Darwin's theory could be considered 'optimistic' is, of course, open to some debate. By 'finalistic optimism' Crocker and Mayer would seem to be referring to the idea that evolution is not blind but has a direction, and indeed a goal, and that this goal essentially amounts to the improvement and perhaps ultimate perfection of humankind. The view of Robinet (see above, p.161 and n.97) might be described as exhibiting 'finalistic optimism'.

into a smaller creature.[133] Nevertheless a contrary idea occasionally comes across as well. Diderot suggested in a number of places that the development of species may be progressive, or at least uni-directional. Nothing less is implied in *L'Interprétation de la nature*, where there is offered what amounts to a (very brief!) history of life and the birth of human civilisation:

> l'embryon formé des ces éléments a passé par une infinité d'organisations, et de développements; qu'il a eu par succession, du mouvement, de la sensation, des idées, de la pensée, de la réflexion, de la conscience, des sentiments, des passions, des signes, des gestes, des sons, des son articulés, une langue, des lois, des sciences, des arts; qu'il s'est écoulé des millions d'années entre chacun de ces développements.[134]

The point is demonstrated even more firmly in the *Rêve* itself, where Diderot made a number of extraordinary suggestions. First, upon contemplating the manner in which organs, needs and habit are connected, Diderot speculated that eventually Man may be nothing more than a head; in other words, that as a species we are evolving towards something that simply encapsulates our best and most commonly used faculty: thought.[135] Several pages later it is suggested that Man is 'une machine qui s'avance à sa perfection par une infinité de développements successifs'.[136] This is not to say of course that Diderot felt mankind would ever be absolutely 'perfect' (in any kind of moral sense); he viewed the world in a fashion too relativistic to allow such universal judgements to be made. Still, this example does provide evidence that there was some consideration of (albeit neutral) directionality in Diderot's proto-evolutionary thought.

What should be made of all of this? Where, in the history of the development of evolutionary thought, should Diderot's work be placed? First, it should be noted that to compare Diderot's ideas too stringently with later, more developed theories of evolution is anachronistic and potentially misleading. Diderot's views are always bound to lose out in any comparison with those of, for instance, Darwin. The goals of the two men were dramatically different. Darwin set out to articulate a comprehensive theory of evolution, while Diderot's speculations were subsumed within works of broader concern. The latter did not attempt to gather empirical evidence in support of his thesis, but merely developed a theory, or more accurately, a hypothesis, on the basis of observations and ideas provided by others. Perhaps more importantly, Darwin's theory of evolution was underpinned by a statistical argument; Diderot's ideas are concerned simply with form, rather than with number. As a result, Diderot's proto-evolutionary thinking lacks what may be regarded as

133. *Le Rêve de D'Alembert*, p.91.
134. *Pensées sur l'interprétation de la nature*, p.769.
135. *Le Rêve de D'Alembert*, p.97.
136. *Le Rêve de D'Alembert*, p.109-10.

one of the key elements of modern empirical science. Last, though certainly not least, the true scope and operation of Diderot's proto-evolutionary thought emerges only upon drawing together scattered references from manifold works written over the course of approximately thirty to thirty-five years (depending upon whether or not one chooses to include *Les Aveugles* in the reckoning).

More importantly however, a strict comparison of Diderot's thoughts on this issue with those of Lamarck or Darwin or Wallace should not be the most important consideration. Many fruitful conclusions can be drawn here simply within the context not only of the mid-eighteenth-century French intellectual landscape, but within the context of Diderot's own work. First, the seeds of evolutionary thought found throughout Diderot's work were sufficiently developed to indicate that he meditated long and deeply on these issues. While not drawn together into a single work, the totality of Diderot's ideas about the development of species is reasonably intelligent and systematic enough to be considered an integrated, coherent hypothesis.

An examination of this proto-evolutionary hypothesis also reveals that Diderot's ideas underwent significant development. Over the course of his career Diderot's thoughts on the issue became increasingly complex, were substantiated with reference to more concrete evidence and were expressed in what may be called a more scientific fashion. *Les Aveugles* presents its reader with a very powerful and intoxicating vision of a world subject to ceaseless upheaval, in which many species are wiped out by the cruel vicissitudes of a godless nature. *L'Interprétation de la nature* may elaborate upon this view somewhat, but still offers up the idea of evolution not as a firm claim but in the form of the statement 'why shouldn't it be the case that...?'. By the time of *Le Rêve de D'Alembert* and especially *Eléments de physiologie* however, the notion has become much more concrete. It is articulated with less wariness, greater confidence and with reference to evidence and ideas that are more scientific. Maupertuis's ideas about heredity, for instance, have been assimilated, as has the notion that the passage of characteristics through generations is governed by a set of laws that are as yet unclear but nevertheless very real.

Likewise, the crucial concept of mutation is added later, or at least acquires increasing prominence over the course of Diderot's career. Early on, he speaks only in terms of 'nature's modifications', but later, Diderot talks about inevitable accidents in the process of generation. The link here with his burgeoning consciousness of heredity is obvious. What may not be so obvious is the possibility that the chemistry course of Rouelle could have provided him with seemingly concrete evidence that organic material, like all material, can be mutated, seemingly of its own accord, even at the molecular level. A sense of this arises in a work that we have not discussed here but that was treated exhaustively in the previous chapter, *Principes philosophiques sur la matière et le mouvement.*

The process of increasing sophistication is also revealed with regard to the progressive aspect of Diderot's notion of evolution. The cursory history of life sketched out in *L'Interprétation de la nature* is almost poetic. Much more unequivocal are his statements in *Le Rêve de D'Alembert* about species developing not towards some final goal, but at least in a fashion that connects development to the faculties and functions of an animal, and which may be seen as an amelioration of a species' condition. Needless to say, his reliance here and in *Eléments de physiologie* on the needs/organs thesis lends Diderot's notion of evolution further depth and sophistication. All of these examples are interesting because they demonstrate that Diderot was assiduous and serious about his natural philosophy. He attempted to keep up to date with all contemporary developments. More significantly he tried to ensure that his own hypotheses were formed and argued in a way that was entirely credible. This is particularly apparent with regard to *Eléments de physiologie*, which Diderot's literary executor Naigeon claimed was an attempt to write a serious, up-to-date, comprehensive scientific tract. As Jean Mayer has noted, it was not a bad attempt either – in fact it stands up quite well to everything else that was available at the time.[137]

Another benefit to have been derived from such a close examination of both Diderot's proto-evolutionary thought and his ideas about *sensibilité universelle* is that it emphasises the value and function of hypotheses in his work. These two aspects of his work are important instances of Diderot's particular manner of melding observational evidence (whether it is gathered by himself or by others), inductive techniques and creative speculation within a framework designed to carefully, defensibly and inevitably (hopefully) advance Man's knowledge of the world – that is, to assist Man in assembling an ever more accurate and sophisticated model of reality, and in so doing enable him to manipulate it better.

Finally, the particular development of these hypotheses demonstrates once again that Diderot's natural philosophical world-view is comprised of a number of layers which are not contradictory. While they may all be ultimately interconnected and have points that overlap, they are, in some senses, strikingly different. Yet this difference does not make any of them less 'true' or useful. Each layer is simply a different way of looking at reality, each represents a different level of abstraction, and the proper investigation of nature requires the employment (albeit to different degrees and with different qualifications attached to each) of all of them. This is made manifest here through Diderot's use of hypothesis – both that of 'sensibilité' and that of evolution – hypotheses that vary from one used in physio-mathematics,[138] but nonetheless very useful in Man's

137. See the introduction to Jean Mayer's critical edition of *Eléments de physiologie*, and also Kiernan, 'Additional reflections', p.131.

138. Compare, for example, to *Sur la cohésion*, vol.5. See also chapters 1 and 2.

pursuit of understanding, or the construction of order (such as it can be in Diderot's universe).

Apart from these remarks about the general nature of Diderot's natural philosophy, we may also now clearly state his response to the question, what is life? Life was for Diderot something most probably inherent in matter. It was only truly manifested or perceptible in beings that were organised in a particular fashion, but Diderot also suspected that it had to rely to some extent on something in the very nature of matter itself; a mysterious principle or vital force, primary and fundamental perhaps to even the most basic laws of physics and chemistry, at least as those laws existed at the time. Despite this, he did suggest or speculate that the sophistication of these laws (law being, properly speaking, not rules which exist in nature but rather human formulations used to explain relationships between observed phenomena) may well lead to a more sophisticated understanding of life and its operations.[139] Diderot also proposed that there existed a certain regularity in the way that life operated: there are, for instance, rules governing the process of generation, even if these are currently unknown. He also posited that the form of beings is determined by both internal and external causes, and more spectacularly, that these forms are not only susceptible to change but are indeed always changing.

Implicit in Diderot's thought is that life, in spite of its mysteriousness, must be understood in conjunction with everything else. It is part of a thoroughly interconnected reality, and its study must be part of the co-operative effort represented by a series of different and yet methodologically linked disciplines. Understanding life is but part of the process of understanding nature; its consideration cannot be separated – even from death. In fact, Diderot posited that life and death themselves are part of the same spectrum. In the *Encyclopédie* article 'Naître', he argued that life is but the successive juxtaposition of particular molecules, and that death is simply the dissipation of these molecules, it is their return to the 'general mass'.[140] Life is held here by Diderot to be an essential quality of the matter that constitutes living beings, and the distinction between 'inert life' and 'active life' has only to do with the removal of obstacles.[141] In *Le Rêve de D'Alembert*, Diderot shows how this removal of obstacles might take place as a marble statue is ground up, made into earth, some of its particles are assimilated into the vegetable matter that grows in the earth, and this vegetable matter is in turn assimilated by a human being who eats it.[142] *Eléments de physiologie* also

139. More will be said on this in the next chapter, as the distinction and interplay between mechanism and (what is traditionally referred to as) vitalism in his thought is only properly nuanced with regard to his view about the nature of human beings.

140. 'Naître', *OC*, vol.15, p.341.

141. 'Naître', p.342.

142. *Le Rêve de D'Alembert*, p.58.

posits that there is no real line between life and death, and that neither state is immediate or absolute.[143]

All that remains then, is to further investigate Diderot's manifold ideas about what he undoubtedly regarded as the most intriguing and important manifestation of life: Man.

143. *Eléments de Physiologie*, Quintili edn, p.129.

5. Medicine

In *Eléments de physiologie* Diderot remarked: 'pas d'hommes dont la conversation soit plus intéressante pour moi, que celle des médecins.'[1] This is an alluring statement, especially in light of Diderot's eclecticism and unusually broad range of passions. To state that medical men were his preferred companions, of all the people he had the opportunity to liaise with throughout his remarkable career, is noble praise indeed. Such admiration and interest begs many questions, the most obvious of which is why Diderot would take such a profound interest in what doctors had to say. Clearly, the discipline was crucial to his natural philosophy, and it played a key role in determining the shape and structure that philosophy was to take.

In some respects the task of unravelling the nature and meaning of Diderot's medicine is more daunting than for any of the other realms, since his knowledge in this area was very wide ranging. Moreover this plethora of knowledge was put to work by Diderot on an equally impressive range of fronts; that which he learned from his esteemed medical friends filtered through into scientific texts, moral philosophy and even literature. Part of the reason for such diversity of both knowledge and its application undoubtedly stems from the fact that Diderot had an enormous variety of medical sources. If Diderot's self-proclaimed fascination with medicine exceeded that of the other disciplines, so too his knowledge of the subject was harvested with the use of a net broader than that used to build his understanding of the other realms. It makes sense therefore to turn our attention first to the nature of Diderot's medical knowledge, and the means by which it was acquired.

The story of Diderot's interest in medical matters begins, perhaps, with his very upbringing. Although it is relatively common knowledge (amongst Diderot scholars, at least) that his father was a cutler, it is less well known that Didier Diderot specialised in the fabrication of surgical implements.[2] The temptation exists here to assert that this fact may have predisposed Diderot to an appreciation of the medical realm, or at the very least that the occupation of his father familiarised him with existence and role of the healing arts. Perhaps even during these earliest of years, the young Diderot came into contact with surgeons, doctors or other medical professionals. Perhaps he took an interest in the tools constructed by his father, pondering or even questioning as to their

1. *Eléments de physiologie*. Quintili edn, p.351.
2. Wilson, *Diderot*, p.12-13.

prospective uses. This is to indulge in pure speculation of course, but it is a fact worth stating, for it is at their peril that intellectual biographers ignore even briefly surveying the milieu of their subject's childhood.

But if pure speculation is all that is possible regarding Diderot's earliest years, it is at least more than is possible regarding medical matters for the subsequent three decades of his life. By the mid 1730s, as we have seen, Diderot had finished with his formal education and was living a bohemian existence in Paris. With the exception of studying Aristotle's *De anima* as part of his philosophy course, his education would not have touched upon medicine, and from Diderot's own account, there is no indication of any interest whatsoever in discipline at this time. His two great loves were classics and mathematics; all other intellectual endeavours were overshadowed by these, and if there did exist an interest on Diderot's part in anything else the lack of evidence means that it must necessarily remain in obscurity.

The first confirmed expression of interest in medicine on Diderot's part was his attendance of César Verdier's privately held course on anatomy during the early 1740s. It is not known how assiduously Diderot attended these sessions, nor for how long. More difficult still is the prospect of establishing what motivated him to learn something about anatomy. The only information available comes from his correspondence, which notes that Diderot frequented a series of lectures on anatomy by César Verdier held at the house of Marie-Catherine Biheron, herself an enthusiast in the field.[3]

The sole publication of César Verdier dealt primarily with muscles and bones, and it is likely that his lectures shared a similar bias. As Jean Mayer has noted, these two elements of the body were amongst those least discussed by Diderot later in his career, from which one might conclude that the Verdier course had little impact on his intellectual development.[4] Nevertheless the significance of what appears to be Diderot's first exposure to the medical realm should not be underestimated. At the very least, Verdier's course would have provided him with a glimpse of the complexities of the human machine, and with a good, general introduction to the structure of the body.

Diderot's attendance of this course was relatively contemporaneous with the commencement of his work on translating Robert James' *Medical dictionary*. Having already worked for the publisher Briasson in producing

3. For evidence of Diderot's attendance, see Roth's edition of his *Correspondance* (vol.8, p.211). Mlle Biheron was well known in mid-eighteenth-century Paris for the quality and accuracy of her wax anatomical models. Her work was highly praised by Diderot, and she even delivered instruction to the *philosophe*'s daughter, Angelique (See Wilson, *Diderot*, p.594-95). For more on Mlle Biheron see P. Dorveaux, 'Notes sur Mlle Biheron', in *La Médecine anecdotique, historique, littéraire* (Paris, Gallimard, 1900-1901), p.165-71.

4. Mayer, *Diderot, homme de science*, p.240.

French versions of Temple Stanyan's *History of Greece* and Shaftesbury's *Essay on merit and virtue*, Diderot was accompanied in the labour of translating James' work by François Vincent-Toussaint, Marc-Antoine Eidous and the young doctor Julien Busson. The *Dictionnaire universel de médecine* was sold by subscription, and appeared in several volumes from 1745 through 1747 inclusive.

The translation, of course, varies from James' version, but not simply in terms of style. The content of the French edition is different from the original in a number of subtle yet significant ways, betraying the interests and concerns of its translators. Perhaps the most striking innovation is an increased emphasis on the thought and work of Hermann Boerhaave, the early-eighteenth-century Dutch physiologist.[5] More generally the Briasson edition is marked by periodic injections of philosophical prejudice. As James Doolittle notes, there are ideas suggested in the French translation that are entirely absent from James' work. Diderot's article 'Acmella' is held up as an example of such alteration, importing what Doolittle describes as 'proto-evolutionary' themes of dynamism and transformism.[6] This in itself opens the door to a vast realm of possibilities. Could it be that in this translation of James' *Medical dictionary* the scholar has access to what may be the seeds of much of Diderot's mature thought?

While there does exist potential in this line of questioning, the fact remains that in the mid 1740s Diderot's medical knowledge was still very much in its infancy. It is extremely unlikely that he was in the position to make significant and substantive alterations to any technical material contained within the work of James. Any such modifications are much more likely to have come from the hand of the trained professional Busson (1717-1781).[7] If articles translated by Diderot himself do contain themes that embellish upon or diverge from that which is in James, it is likely that the source of such liberties does not rest so much in actual medical knowledge as in more general philosophical biases. One possibility that pertains directly to the aforementioned article is that Diderot's amendments were inspired by his work on Shaftesbury's *Essay on merit and virtue*. This work contains many of the same themes as the article 'Acmella' and,

5. Catherine Raymond, 'Diderot et la médecine', in *Litterature et médecine II*, ed. Jean-Louis Cabanes, *Cahiers du laboratoire pluridisciplinaire de Recherches sur l'imaginaire appliquées à la litterature*, no. 55 (July 2000), p.216. Boerhaave was head of the Leiden University medical faculty in the early eighteenth century, and was one of the era's most influential proponents of iatromechanism. For more on Boerhaave and his theoretical outlook, see chapter 3 n.18 and p.198-99 below.

6. James Doolittle, 'Robert James, Diderot, and the Encyclopédie', in *Modern language notes* 71 (June 1956), p.432-33.

7. A native of Brittany, Julien Busson spent the early years of his career as a Parisian doctor and as physician to the Duchesse du Maine. He later attained some measure of notability on account of his holding several important administrative posts, such as physician to the commandant en chef, the duc d'Aiguillon.

in both cases, these themes are dealt with not in a technical way, but in a philosophical or even metaphysical manner.[8]

Like the Verdier course then, Diderot's work on translating James would have served primarily as a learning experience. From it, he would have acquired a very broad base of knowledge, one that extended far beyond the confines of anything that Verdier taught. All aspects of medicine were covered in James' work and, just as importantly, many ways of thinking about medicine. It is also worth noting that from the experience of translating the *Medical dictionary* Diderot may have taken away many ideas regarding the interconnection of various branches of knowledge, as well as ways of organising a broad expanse of material in an encyclopedic format. These lessons would obviously benefit Diderot tremendously when it came to assembling the *Encyclopédie*.

Perhaps most importantly, however, Diderot's tenure as translator of James seems to have sparked or at least accelerated his interest in the medical field. Immediately subsequent to his work for Briasson he began to publish pieces that dealt with medical matters. The most obvious example is *Lettre d'un citoyen zélé*, which originally appeared in 1748.[9] Broaching the issue of how to reconcile the professional differences that existed between physicians and surgeons, the work highlights Diderot's new-found concern with the medical professions. The ramifications of Diderot's argument shall be treated in greater detail shortly – what matters here is that the work's very composition is indicative of the rise in the latter half of the 1740s of Diderot's appreciation of issues related to the medical world.

Another example of this burgeoning interest is *Lettre sur les aveugles*, also of 1748. This work discusses a wide variety of issues, including God, cosmology, the arithmetic machine of the English mathematician Saunderson and, of course, the nature and implications of blindness. Diderot's treatment of the latter issue is remarkable both for its citation of a number of experiences related by doctors and surgeons, and for its role as his first genuine attempt to come to terms with the human senses and the manner in which they govern Man's knowledge of reality. Present here is Diderot's nascent fascination with psychology; very embryonic at this point, but significant nonetheless. It may be argued that the standpoint from which Diderot addresses the issues here is more philosophical or even metaphysical than it is technical and medical but, as will become increasingly apparent, medicine and philosophy were to a large degree inseparable both for Diderot and his contemporaries. Moreover it was to medicine that Diderot turned in order to gain a firmer understanding of

8. Themes, for instance, pertaining to the dynamic, organic and fundamentally unified nature of reality, to time and interconnectedness. See Schlegel, *Shaftesbury and the French deists*, p.43-62.

9. Chapter 3 of this book has of course already discussed the nature and meaning of the revisions that were carried out to the *Lettre* for its 1760 edition.

the problems discussed in *Lettre sur les aveugles*. It was at this time that Diderot, rather than simply relating the experiences of doctors (or even of natural philosophers carrying out investigations) second hand, managed to gain access to a number of cataract operations, one performed by Réaumur and one witnessed while imprisoned at Vincennes.[10] This was just the beginning of what would be for Diderot a long career of playing spectator to the medical profession. Throughout his life he would witness many instances of medical men at work, be they surgeons at the operating table, or doctors in clinical situations.

Probably the most important of the medical men that Diderot was fortunate enough to witness 'in action' was Théophile de Bordeu. Diderot met Bordeu through Julien Busson and, despite the breadth of knowledge acquired through translating James' *Medical dictionary*, meeting Bordeu may in fact have been the most important consequence of his affiliation with that enterprise.[11] Bordeu was a key figure in the medical world of mid-eighteenth-century Paris, although he was originally a member of the medical faculty at the University of Montpellier.[12] The faculty of this provincial university, as opposed to that of Paris, tended to foster more radical approaches to medicine; it more often than not preceded the capital in terms of adopting and promoting challenges to prevailing medical orthodoxy.[13] Bordeu was not an exception to this Montpellier tradition. Following in the footsteps of Boissier du Sauvages, he became one of eighteenth-century France's most important promoters of medical vitalism, a school of thought which denied the proposition that the phenomena of life were reducible to the simple mechanical operation of bodily parts, maintaining instead that living beings were directed by principles different in kind from those that governed the inanimate world.[14]

10. There is of course a certain poetic irony in this, for as will be recalled from chapter 4, acrimony stemming from Diderot's attendance at the cataract operation performed by Réaumur may have been part of the reason that he found himself landed in prison at Vincennes shortly thereafter.

11. Paitre, *Diderot biologiste*, p.19.

12. Hailing from Montpellier and working in Paris were not, of course, mutually exclusive. From as early as the seventeenth century Montpellier physicians were being licensed to practise in the capital, and indeed there had existed a long tradition of doctors educated at the Faculty in Montpellier serving as chief physician to the King. See Roger, *Buffon*, p.48.

13. Montpellier had, for example, introduced the mineral-based pharmacopoeia of Paracelsus to France in the seventeenth century (see Roger, *Buffon*, p.48). It was also the prime point of entry into France for the vitalistic doctrine of Stahl (see below p.199). For more on the role of the Montpellier medical faculty in shaping the course of eighteenth-century French thought see Brockliss and Jones, *Medical world of early modern France*.

14. Of course, as we shall see shortly, Bordeu's vitalism differed from that of Boissier du Sauvages in a number of key respects (see below, p.199-200). For a nice summation of Bordeu's work and thought, particularly within the context of the 'Montpellier School', see Elizabeth Williams, *The Physical and the moral; anthropology, physiology, and philosophical medicine in France, 1750-1850* (Cambridge, 1994), p.32-41.

It has been suggested that most Montpellier physicians tended to share a bias against extensive systematisation in favour of a more empirical approach to medical research, and indeed part of the reason for Bordeu's rise to fame was on account of a number of important experiments published in the early 1750s.[15] The most significant of his researches centred around the means by which glands secreted their various fluids (*Recherches anatomiques sur la fonction des glandes et sur leur action*, 1751). Through a series of well-construed trials Bordeu demonstrated that the process by which glands secreted fluid was not due to compression, as had been supposed by most mechanists.[16] For Bordeu, glandular activity was to be explained in terms of 'a kind of convulsion', which itself was the result of nervous action.[17] This denial of simplistic, mechanical explanations for basic bodily functions was indeed part of the reason for Bordeu's support of vitalism. While not necessarily confirming the existence of some vital force that impelled organic matter to function as it did, Bordeu's experiments on glands highlighted the inadequacies of conceiving of the body as 'hydraulic machine', and therefore lent weight to the drive for more open-minded approaches to defining and describing life and the body.[18]

Bordeu also sketched out a complex and novel physiology, which, as we shall see, exerted considerable influence on Diderot's conception of the human being. Bordeu viewed the body as a more-or-less loose assembly of organs, hierarchically arranged, in which each vital organ possessed a certain degree of autonomy. This autonomy was conceived in terms of the particular organ's nature, role, needs and individual state of health. Human life was, for Bordeu, a multi-layered phenomenon. The life of the body as a whole was superimposed over the lives of the organs themselves, which could be viewed as semi-independent systems in their own right. Thus, according to Bordeu's theory, the unity of an individual might be conceived as a Venn diagram, in which each of the organs that constitute the body can be looked upon as autonomous circles which may or may not interlock at certain points, the entirety of the diagram representing the total self.

15. On the generally empirical stance advocated by most members of the Montpellier school see Roger, *Life sciences in eighteenth-century France*, p.535. It should be noted that an alternate view (asserting that Montpellier physicians tended to share a bias against experimentation) has been advanced by Martin Staum in his book *Cabanis; Enlightenment and medical philosophy in the French Revolution* (Princeton, NY, 1980), p.309.

16. Using a cadaver, Bordeu inserted a wet sponge in the cavity occupied by the parotid gland and then worked the jawbone as in chewing. Finding that the sponge was not compressed, he concluded that the mechanical operation of parts surrounding the gland could not account for its secretion of fluids. See Williams, *The Physical and the moral*, p.34.

17. Williams, *The Physical and the moral*, p.34.

18. More on Bordeu's research can be found in Haigh, *Xavier Bichat and medical theory*, p.32-36.

In terms of actual medical practice, one of the most notable aspects of Bordeu's thought was his consummate respect for nature and for nature's capacity to fix whatever may be wrong with a particular body. His therapeutic technique, therefore, tended to be very non-invasive. Rather than resorting to pharmacy or worse yet surgery, Bordeu preferred to let nature do the work of healing the patient – the role of the doctor was regarded primarily as that of giving nature assistance when necessary. These points are made manifest in the *Encyclopédie* article 'Crise', in which Bordeu argues that the employment of harsh purgatives and bleeding should be avoided whenever possible. The article indicates that when action is necessary, preference should be given to milder solutions such as the use of laxatives and that doses of medication should generally be smaller than those usually prescribed by doctors. Thus Bordeu did not rule out intervention out altogether – it would be necessary from time to time – but his approach to taking such measures was cautious and relied on waiting for nature to indicate to the physician when action should be taken.[19]

Much as he did with G. F. Rouelle, Diderot enjoyed a close and long-lasting relationship with Bordeu, one that bridged the gap between professional and personal. As was also the case with Rouelle, this relationship was not entirely one way; it was not simply a case of Diderot benefiting from the experience and wisdom of one who was an acclaimed master in a given field. Elizabeth Haigh suggests that Bordeu himself reaped rich rewards from his association with Diderot, and that a close relationship with the gregarious and well-connected *philosophe* helped the doctor's reputation to grow in Paris.[20] It was also through Diderot that Bordeu was able to contribute a number of articles to the *Encyclopédie*, including the aforementioned article 'Crise'. Of course, in this particular instance, it is entirely unsure who may have been performing the favour!

The personal closeness of Diderot and Bordeu is affirmed by the fact that, over a period of several years, the renowned physician attended to many people near to Diderot. An inspection of Bordeu's correspondence reveals that he doctored the mother of Sophie Volland, her sister Damilaville (Diderot's most frequent correspondent), Mme Diderot and Julie Lespinasse.[21] On several of these occasions Diderot was present, giving him ample opportunity to witness one of the century's most renowned physicians conducting diagnoses, prognoses and therapies.

Diderot also went out of his way to visit Bordeu and questioned the doctor incessantly about various matters related to medicine. We can even speculate that after a time Diderot almost became something of a

19. Théophile de Bordeu, 'Crise', *Encyclopédie*, vol.4, p.481.

20. Haigh, *Xavier Bichat and medical theory*, p.32.

21. See *Théophile de Bordeu: Correspondance*, 3 vols, ed. Martha Fletcher *et al.* (Montpellier, 1977-1979).

pest, rather like the child that wants to know too much, for eventually Bordeu simply wound up by furnishing Diderot with a parcel of technical notes.[22] It is slightly endearing to imagine the scene: Diderot going over to Bordeu's full of questions about physiology, medical research and clinical technique, only to have the doctor sigh with exasperation and hand the over-curious philosopher a large bundle of papers saying 'take and read!'. But whatever the exact nature of their transactions – personal, philosophical, scientific – there is little doubt that Diderot learned a great deal about medicine from Bordeu. Indeed, such was the mass of medical knowledge acquired by Diderot that Bordeu referred to him as 'médecin Diderot'.[23]

While Bordeu was the most important figure in Diderot's medical education, he was not the only one from whom Diderot learned something. The *Encyclopédie* was a prime source of contact with doctors. Although it is not always possible to determine the degree to which Diderot was involved with any of them personally, it may be presumed at the least that he became familiar with much of their work. A list of those to whom he would have been exposed includes Fouquet, Menuret de Chambaud, Venel, Tronchin and Le Roy.[24] Many of these contributors were of the Montpellier Faculty, a fact which itself is worthy of further investigation. Could it be that this Montpellierian coterie was assembled more or less by accident, or perhaps under the influence of Bordeu, and that it subsequently exerted an influence on Diderot's perception of medicine? Or might it have been the other way around? Might Diderot have already harboured preferences for those who thought along lines similar to Bordeu, and therefore purposefully contracted many Montpellierians to the enterprise in order to lend the work particular overtones?[25] There is, of course, danger in trying to overplay this relationship in either direction, as the *Encyclopédie* received contributions from a wide variety of quarters. A perfect demonstration of this is the fact that the project's most prolific writer, the Chevalier de Jaucourt, studied medicine under Boerhaave in Leiden and (though not as dogmatic as his mentor) did possess some distinctly mechanist leanings.[26] Instead of

22. Paitre, *Diderot biologiste*, p.84.

23. Raymond, 'Diderot et la médecine', p.221.

24. More on the respective careers and contributions of these men can be found in Frank A. Kafker, *The Encyclopedists as a group: a collective biography of the authors of the Encyclopédie*, *SVEC* 345 (1996).

25. These questions require further research. Although Jacques Proust's *Diderot et l'Encyclopédie* does offer some insight as to the manner in which the encyclopedists were assembled – even analysing the nature of their respective backgrounds and social status – he offers little elaboration in the way of medicine. See Proust, *Diderot et l'Encyclopédie* (Paris, 1962).

26. On Jaucourt's contribution to the *Encyclopédie*, see John Lough, *The Encyclopédie* (London, 1971), p.44-45 and p.73. On his tempered mechanist approach to medical theory, see Brockliss and Jones, *The Medical world of early modern France*, p.425 and p.437.

emphasising Diderot's connections with a particular school of medical thought, it therefore makes more sense to acknowledge the multiplicity of sources from which he obtained medical knowledge. These sources were as varied in medium as they were in philosophical bias. Beginning in the 1750s, for example, and continuing until the end of his life, Diderot frequented a number of important salons – from the polite gatherings at the home of Julie Lespinasse to the more radical ones at that of Baron d'Holbach. Each of these forums provided the opportunity to mix with doctors and surgeons of every description.[27]

Diderot's correspondence provides further insight into the manner in which he expanded his knowledge of medicine. Particularly during his later years, there is frequent correspondence with a number of medical men. Antoine Petit (1718-1794) was questioned by Diderot in 1771 regarding matters of physiology, and the answers received contributed to the assembly of *Eléments de physiologie*.[28] Another correspondent was the Dutch physician and naturalist Petrus Camper (1722-1789), whose study on human facial angles was much admired by Diderot.[29] After returning from his trip to Russia, Diderot addressed the short work *Observations et questions sur la variole* to Camper, hoping to obtain greater understanding of the history and nature of smallpox. Other doctors and surgeons who appear in Diderot's correspondence include Jacques Dupau and the aforementioned Le Roy, to whom the *philosophe* addressed questions regarding animal heat.[30]

In addition to meeting and corresponding with a number of medical professionals, Diderot read very widely on the subject of the human body. Though, as usual, there is little indication of exactly when he may have read any given book, it can be confirmed that by the time he was composing *Eléments de physiologie* (a process that extended throughout much of the 1770s) he had familiarised himself with the thought of Barthez (*Nouveaux éléments de la science de l'homme*, 1778), Boerhaave, Bonnet (*Essai analytique sur les facultés sur l'âme*, 1760), Borelli, Cullen (*Institutions of medicine*, 1772), Cureau de la Chambre, Fontana, Galen, Haller (*Elementa physiologiae*, 1766), Heister, Helvétius (*De l'esprit*, 1758 and *De l'homme*, 1773), Hippocrates, La Mettrie (*L'Homme machine*, 1746;

27. On the salon of Julie de Lespinasse, see Janine Bouissounouse, *Julie de Lespinasse* (Paris, 1958). On the gatherings that took place at the baron d'Holbach's estate Grandval, see Haigh, *Xavier Bichat and medical theory*, p.75, Paitre, *Diderot biologiste*, p.8, and Alan Kors, *D'Holbach's coterie: an enlightenment in Paris* (Princeton, NJ, 1976).

28. *Correspondance*, vol.11, p.70. Petit occupied the chair of anatomy at the Jardin du Roi and was one of the most successful and popular lecturers on medical subjects in the third quarter of the eighteenth century (see Brockliss and Jones, *Medical world of early modern France*, p.450).

29. Wilson, *Diderot*, p.654. Wilson also suggests that Diderot and Camper enjoyed at least two very amiable meetings, one in Paris, the other in the Hague (see Wilson, p.689).

30. *Correspondance*, vol.15, p.138.

and *L'Homme plante*, 1749), Le Camus (*Médecine de l'esprit*, 1753), Marat (*De l'homme*, 1775-1776), Peyrilhe (*L'Histoire de la chirurgie*, 1780), Roussel, Spigelius, Stahl (*Theoria medica vera*, 1708), Sydenham, Whytt (*Essay on the vital and other involutary motions of animals*, 1751) and Zimmermann (*Traité d l'expérience*, 1763).[31] Represented here, just as in the list of contributors to the *Encyclopédie*, are all the major schools of eighteenth-century medical thought, and more. Diderot exposed himself to everything from the Classics (on whom the bulk of scholastic medicine was based until the late seventeenth century) to the most contemporary thinkers; to works by French, English, Dutch, Swiss and Italian authors; to the most die-hard iatromechanists and the most animistic of vitalists.

Thus, despite a relatively late start, Diderot had by the height of his intellectual career obtained an extensive and well-rounded knowledge of medicine. This knowledge was the product not only of book learning but of conversation and correspondence with some of the most renowned medical thinkers of the day, and of direct observation of doctors and surgeons practising their respective arts. The process by which Diderot acquired his medical 'education' was probably more informal and unstructured, yet on the whole more direct, than was the case with any other branch of his natural philosophical learning. He became, by the 1750s at the latest, firmly entrenched within the medical world of eighteenth-century France; a bona fide member of the coterie of physicians and surgeons who dictated the parameters and direction of medical discourse. Diderot's own experience of this culture may have been dominated by physicians with connections to Montpellier, but this influence was certainly not exhaustive.

The more intriguing problem that arises now is determining exactly what Diderot did with all of this information. To pose the question in such a manner is, of course, somewhat artificial, for Diderot's medical knowledge, as with all other branches of his natural philosophy, was organic; it grew, developed, and took shape over time. Moreover the doctors from whom Diderot gleaned medical knowledge were not simply providing him with evidence, but were guiding his opinions at the same time; they would have been providing not only answers but many of the questions as well.

It is equally important to note however that the relationship between Diderot and doctors should not be configured simply as a one-way street, with the physicians as teachers and Diderot as the pupil. Diderot felt that there was a great deal that he wanted to know pertaining to the human body, and he perceived that doctors could provide him with much useful

31. The bulk of these works, some of which date from the seventeenth century, constitute the bibliography of *Eléments de physiologie* (See Jean Mayer's critical edition, p.xiii-xiv). Boerhaave, Galen, Hippocrates and Sydenham were recommended by Diderot for pedagogical purposes in *Plan d'une université*.

information, but things were not necessarily as straightforward as this. Although, as was suggested above, doctors would have been furnishing Diderot with many of the questions as well as some of the answers, Diderot was undoubtedly guided in terms of what he wanted to know by his own preconceptions, prejudices and philosophical demands. In other words, the *philosophe* was never approaching medicine from an entirely neutral perspective in the first place – the entire story of his participation in the 'medical coterie' of mid-eighteenth-century France was played out against the background of his already developing natural philosophical world-view. He was tackling the entire issue of the human body with the benefit of already having much knowledge of subjects such as physics, natural history and, after 1754, chemistry. More significantly perhaps, he was approaching medicine as part of an attempt to integrate it within a broader system of ideas. We may, therefore, regard Diderot here less as an apprentice, and more as an intellectual equal of the men he was engaging. Indeed, there may have even existed the possibility that doctors had as much to gain from the relationship as the *philosophe*; that his more well-rounded philosophical perspective or basis of concern lent a novel element to the discourse. These considerations must always provide a sense of perspective when contemplating the medical aspects of Diderot's world-view. Nevertheless it is still legitimate and necessary to ask what Diderot did with medicine. This is fundamental to understanding why it was so important to him, and why doctors were his favourite people with whom to speak.

First, medicine was for Diderot an eminently practical discipline. It was arguably more important from a utilitarian perspective than any of the other branches of natural philosophy. Indeed, Diderot tended to emphasise the use-value of every science that he discussed; he was unambiguous in stating the importance of mathematics and mechanics to erecting buildings capable of standing up, or of chemistry to informing the work of miners, metallurgists and pharmacists. His statements regarding the importance of medicine however are the strongest of all. Diderot's works are rife with testimony as to the supreme significance of fostering good health. This spirit is best captured in *Plan d'une université*, which states that 'Si les hommes sont pauvres, le souverain ne protège que des malheureux; s'ils sont valétudinaires, il ne garde que des malades'.[32]

Jean Mayer asserts that there are two basic conceptions of medicine: one pertaining to the conservation of individual human beings, and the other to the promotion of public health.[33] Mayer maintains that Diderot was primarily interested in the second of these, and this assertion would seem to be supported by the above extract from *Plan d'une université*. Diderot was, however, concerned with both conceptions. A healthy population cannot, after all, be achieved without attending to the welfare

32. *Plan d'une université*, p.817-18.
33. Mayer, *Diderot, homme de science*, p.86.

of individual people. The first manner of thinking about medicine was also paramount in discussing the discipline as a science, as shall be seen in due course. But if Diderot's emphasis on the second ideal was ultimately no greater overall, it was certainly more *à propos* with regard to his thoughts on the use-value of medicine. Health, for Diderot and most of his contemporaries, was a profoundly social issue. The health and welfare of a nation's citizenry was perceived as being very much in the interests of the enlightened state. There is then a strong sense of paternalism (or maternalism in the case of Catherine the Great, to whom Diderot harped upon the practical importance of medicine) connected to the manner in which Diderot outlined the relevance of medicine. This paternalism was even carried by means of analogy to the family, where it would usually be the father who dominated a household: 'Un malade incurable au centre d'une famille, est comme un arbre mort au centre du jardin.'[34]

Public health was for Diderot one of the most important aspects of medicine. Testament to this is the first sentence of Diderot's introduction to the faculty of medicine in *Plan d'une université*, which reads: 'Si l'on veut que des étudiants reçoivent dans une faculté de médecine toute l'instruction qui leur est nécessaire pour exercer l'art de guérir d'une manière utile à leurs concitoyens, il faut se rappeler que la santé publique est peut-être le plus important de tous les objets.'[35] In this context, matters of hygiene and sanitation come to be of paramount importance. Attention to environment and deeper understanding of how contagious diseases are spread form a crucial part of a doctor's role in society. Diderot infers that the medical profession is more relevant to urban life precisely because it is in the cities where levels of health are poorest and where people are most susceptible to contracting disease.

None of this was particularly remarkable within the context of eighteenth-century medicine. Indeed James Riley has shown that an increase in attention to matters of health and environment was part of a general hippocratic revival taking place at the time.[36] Although accomplishments were modest and governmental officials were often slow to undertake activities such as the compilation of medical statistics, there was a growing conviction amongst eighteenth-century physicians that epidemics might be avoidable, and some progress was made in terms of improving drainage, lavation, ventilation and interment.[37]

But of course public health consists not just in safeguarding a given

34. *Plan d'une université*, p.820.

35. *Plan d'une université*, p.817. We must consider here whether or not Diderot is using the term 'santé publique' in precisely the same fashion that we would use it today. Is he referring to the overall well-being of the population, or merely to rectifying certain aspects of hygiene and sanitation? His usage is somewhat ambiguous, and seems to vacillate between the two.

36. James C. Riley, *The Eighteenth-century campaign to avoid disease* (London, 1987), p.ix-x.

37. Riley, *Eighteenth-century campaign*, p.111.

population against disease, but in trying to affect remedies as well. To this end, Diderot promoted the value of hospitals.[38] This promotion must be qualified however, in that more often than not Diderot spoke in terms of the potential of hospitals to improve people's lives, rather than of the reality of their actually doing so. Indeed Diderot was quite pessimistic about the state of hospitals as they existed in the eighteenth century, and was quite sceptical about the benefits that were actually conferred by such institutions. In the *Encyclopédie* article 'Hôtel-Dieu', Diderot suggested that in their current state hospitals were probably better places to get sick than to be healed![39] Nevertheless Diderot was an optimist; he believed that hospitals could and would get better, and that the general welfare of the nation would improve along with them. If such a view seems commonsensical today, it should be noted that in the eighteenth century the role and value of hospitals was a matter of some debate, and there were those who thought that public health would be better served by the 'dehospitalisation' of society.[40] The fact that there existed some measure of public uncertainty regarding the merits of hospitals, in conjunction with the relatively poor record exhibited by such institutions, indicates that in maintaining a fundamentally positive attitude Diderot was expressing a certain amount of faith, probably shared with most members of the Republic of Letters, that the rational progress of medicine, in conjunction with the rational provision of health care measures on the part of the state, would eventually be able to achieve substantial goals in the way of ameliorating the condition of Man in society.

The great eighteenth-century debate over inoculation is another example of this presumption. In a memoir directed against D'Alembert's reservations about inoculating for smallpox, Diderot applied probability theory to the problem of deciding whether or not it is best to conduct mass inoculations.[41] Diderot distinguished between three different ways in which lives of people could be regarded. First there is the 'vie réelle', which considers the entire duration of a life, from the moment of birth to that of death. Next is the 'vie moyenne', or the average lifespan of a member of a population. Finally, Diderot presented the notion of 'vie civile', which is that span of a person's 'real life' during which they are

38. It needs to be stressed here that eighteenth-century hospitals were very different from modern institutions of the same name. To a large extent, hospitals in Diderot's day served as a combination poorhouse, insane asylum, and general place of refuge for the sick, the aged, the young and the dispossessed; they were not exactly regarded as houses of healing. Indeed anyone with money in eighteenth-century France would not have even considered darkening the doors of such a place. For more on the nature, state, and function of the eighteenth-century hospital, see Brockliss and Jones, *Medical world of early modern France*, p.671-729.

39. Raymond, 'Diderot et la médecine', p.218.

40. Brockliss and Jones, *Medical world of early modern France*, p.30.

41. *Nouveaux mémoires sur différents sujets de mathématiques; De l'Inoculation*, *OC*, vol.5, p.41-42.

able to perform functions genuinely useful to society. On the basis of these distinctions, and given the odds for increasing the average lifespan of the population with the use of inoculation, Diderot concluded that despite the risks presented to any given individual, it makes sense for society as a whole to proceed with inoculation.[42] This argument is Diderot at his most utilitarian, and it underlines the supreme value he attached to medicine as a practical concern of the enlightened state.

The memoir on inoculation also demonstrates a very interesting confluence between medicine and mathematics. In approaching the issue largely from the perspective of statistics (the calculated odds of dying from smallpox on account of inoculation as opposed to dying from it naturally are confidently stated at 300 to 7), Diderot was indulging in what may be construed as an embryonic form of modern social science.[43] The work of Diderot, D'Alembert, Daniel Bernoulli and others here was prototypical; it was a foreshadowing of the direction that such discussions would increasingly take in the century to come.[44]

Another major Diderotian theme relating to the practical aspects of medicine was the organisation of the profession, which was again seen largely as the concern and responsibility of the state. His first attempt to broach this issue came with the aforementioned *Lettre d'un citoyen zélé*. Chapter 3 has already demonstrated that one of the major points made here, albeit in the 1760 addition to the work, was that medical men should receive a thorough grounding in chemistry. More than this however, Diderot's *Citoyen zélé* stands out primarily as an attack on the corporate organisation of the medical professions. In characteristically pragmatic spirit, Diderot portrayed the structure according to which the healing arts were organised as being governed by self-interest and vanity, much to the detriment of the professions themselves and, most significantly, to the detriment of patients. The crux of his argument is that the harmful rivalry between the medical and surgical professions should be swept away, and that these two arms of what is essentially one task should be more unified in both theory and practice.[45] Even more interesting perhaps is the fact that the medical profession subjected to the greatest

42. *De l'Inoculation*, p.41-42.

43. Diderot does not state where he gets his raw data from, but we might suspect that it was Bernoulli, if only on account of the fact that Diderot's position regarding the inoculation debate was remarkably similar to that of the famous Swiss mathematician. On Bernoulli, see chapter 1, n.28.

44. For more on the rise of statistics and the science of society during the latter half of the eighteenth century, see Baker, *Condorcet*. On the debates concerning the practice of inoculation itself, see G. Miller, *The Adoption of inoculation for smallpox in England and France* (Philadelphia, PA, 1957).

45. It is interesting to note that the *Encyclopédie* article 'Chirugie', by Antoine Louis, adopts a very similar point of view regarding the nature of the two professions. Louis argues that the task of understanding and healing bodies can be regarded as a single enterprise, with medicine simply existing as the branch that attends to internal matters and surgery the one that administers to the external. See the *Encyclopédie*, vol.3, p.351.

criticism here is that of the physicians. The changes advocated by Diderot required the greatest degree of compromise on the part of those at the very top of the medical hierarchy. According to *Citoyen zélé*, everyone concerned would be much better served if doctors read some texts on surgery, and perhaps even deigned to pick up the scalpel once in a while.[46] A similar vein of criticism is repeated in *Plan d'une université*, in which Diderot lamented that too often the practice of medicine becomes wrapped up with matters of reputation and fortune.[47] In a passage strikingly reminiscent of twenty-first-century concerns, Diderot also wondered how a doctor could visit fifty or sixty patients per day, and wind up treating any of them well.[48]

But despite these thinly veiled condemnations, Diderot's admiration for doctors only really grew with time. Whether this burgeoning respect was due to an increasing personal association with doctors (as opposed to surgeons) over the decades following *Citoyen zélé* can only be guessed at, but by relatively late in life Diderot's enthusiasm for them was very high. *Plan d'une université* states that of the four academic faculties, medicine is the one that is in the best shape. Despite the criticisms mentioned above, his treatment of the state of contemporary medical education was really quite charitable. His main prescription for the improvement of the profession was simply that the training of doctors be made more practical. This emphasis on practicality was applied by Diderot even to the recommended textbooks. Diderot purported that modern writers had a habit of speculating too much, and that this tendency should be tempered with doses of ancient empiricism (hence Galen and Hippocrates standing alongside Boerhaave and Sydenham).[49] Hands-on experience was also valorised, and in the interests of this Diderot suggested the establishment of teaching hospitals where doctors in training could practice their art under the supervision of those more experienced.[50]

46. *Lettre d'un citoyen zélé*, p.147.

47. *Plan d'une université*, p.819.

48. *Plan d'une université*, p.819. It must be noted that these concerns were not unique. Just as Diderot's ideas about the value of hospitals were part of a wider eighteenth-century debate surrounding the efficacy of these institutions, so too his comments regarding doctors were but part of a wider dialogue. It was not at all uncommon for doctors in the early modern era to criticise one another for treating patients in a fashion altogether too cursory. See Brockliss and Jones, *Medical world of early modern France*, p.298-99.

49. *Plan d'une université*, p.825. Modern readers may find this adulation of the Ancients at the expense of the Moderns a somewhat bizarre turning of the tables. This would be particularly so in the case of Galen, the very study of whom was, for most of the early modern era, synonymous with speculative, scholastic, non-empirical learning. We might speculate that Diderot was attracted to Galen's materialism, although there is little concrete evidence to bear this out. As far as Hippocrates is concerned, it should be noted that the ancient empiricist was held in very high regard by most eighteenth-century physicians (referring to Sydenham as the 'English Hippocrates' was probably the highest compliment that his contemporaries could have paid him).

50. *Plan d'une université*, p.820.

Of course, Diderot's views on the state of medicine as a profession and its supreme importance to the welfare of society are but an aspect of his interest in the discipline. Addressing medicine as a practical enterprise also involved some thought about what constitutes the 'art of healing'. Diderot's views here were shaped to a large degree by Bordeu's faith in the prudence of nature and nature's ability to heal.[51] The relatively mild approach espoused by the Montpellierian became for Diderot the epitome of how doctors were supposed to treat patients. Certainly Diderot's presence during the treatment of several of his friends and relations by Bordeu meant that he had many opportunities to witness this clinical style. This cautious, minimal-intervention method is recreated in fiction by the *philosophe* in *Le Rêve de D'Alembert*, when Diderot has the character Bordeu attend to a 'disturbed' D'Alembert:

> Bordeu [to Mlle de Lespinasse]: Est-ce qu'il est malade?
> Lespinasse: Je le crains; il a eu la nuit la plus agitée.
> Bordeu: Est-il éveille?
> Lespinasse: Pas encore.
> Bordeu, *après s'être approché du lit de D'Alembert; et lui avoir tâté le pouls et la peau*: Ce ne sera rien.
> Lespinasse: Vous croyez.
> Bordeu: J'en réponds. Le pouls est bon [...] un peu faible [...] la peau moite [...] la respiration facile.
> Lespinasse: N'y a-t-il rien à lui faire?
> Bordeu: Rien.
> Lespinasse: Tant mieux; car il déteste les remèdes.
> Bordeu: Et moi aussi. Qu'a-t-il mangé à souper?[52]

In addition to highlighting Bordeu's general reluctance to intervene here, it should be noted that Diderot also picked up on his friend's attentiveness to the feel of the skin, the nature of the breathing, and especially his (rather Galenic) concern with the pulse (a subject in which Bordeu was extremely interested).[53]

Despite any of his sympathy for or admiration of surgeons, Diderot regarded their art as a last resort approach to the solution of medical problems. He even spoke out against aggressive, invasive procedures such as bleeding.[54] The key to healing, in Diderot's opinion, lay in careful observation of the patient, proper diagnosis, followed by, if possible, a therapy that focused on letting nature itself do the healing. In light of this, Diderot was impressed by and concerned with matters of hygiene, diet (Bordeu's prescription of goat's milk was regarded by Diderot with extreme fondness)[55] and an understanding of the relationship between a

51. Mayer, *Diderot, homme de science*, p.384.
52. *Le Rêve de D'Alembert*, p.77-78.
53. Brockliss and Jones, *Medical world of early modern France*, p.566.
54. 'Lettre à Sophie Volland', 9 September 1767, in *OC*, vol.7, p.557-63.
55. Raymond, 'Diderot et la médecine', p.218.

patient and their environment. Indeed, his conception as to how best to treat a sick individual was very much a mirror of his ideas about public health and the role of medicine in society as a whole. Much of medicine, according to Diderot, depended upon a delicate interplay between an individual and his surroundings. This concern with the environment is demonstrated very well in *Observations et questions sur la variole*, which lingers on the issue of how disease is related to prevailing climatic conditions.

This neo-hippocratic preference (shared not only with Bordeu, but with many eighteenth-century doctors and surgeons)[56] for a gentle, 'hands-off' approach to medicine betrays two important points. First, it highlights the fact that medicine was not simply about focusing on the body as an object to be repaired, but that it involved adopting an integrated approach to a person's entire life. Medicine was not simply about pharmacy and surgery, but about understanding and regulating for the better all aspects of a person's life and environment – in times of good health and in periods of illness. The title of the first section of *Voyage en Hollande* alludes to this profound connection between medicine and environment: it is called 'Médecine, ou du Pays'. The chapter makes clear that all physical aspects of a country – its air, its water, even the condition and quality of the animals that provide food – are the concern of the intelligent, enlightened doctor. Without taking an interest in a person's surroundings and lifestyle, the doctor cannot best serve his patient. That Diderot clearly perceived the connection between health and environment is further demonstrated in *Réfutation d'Helvétius*, although, as shall be seen in due course, this is done with an even more philosophical point in mind.

The second major issue to arise here is that Man is an inextricable part of nature. This lesson is, of course, one that has been addressed in earlier chapters, but it becomes perhaps only fully relevant when contemplated in the context of medicine. By following Bordeu's essential belief that the best healer is nature itself, Diderot was reaffirming his well-established view that Man is part of nature, that he can never be removed from it, and that from a purely physical perspective, Man can and must be regarded as just another animal. In other words, according to Diderot's perception, Man is objectified and materialised. The processes which govern the operation of all organic matter also govern human beings. It is therefore nature as a whole which takes precedence: Man must be regarded as dependent upon and superseded by nature. It is in this spirit that Diderot has the fictional Bordeu of *Le Rêve* state: 'La nature ne

56. Antoine Louis, who authored the article 'Chirurgie' for the *Encyclopédie*, readily accepted that surgery should only ever really be a last resort. See 'Chirurgie', *Encyclopédie*, vol.3, p.350-55.

souffre rien d'inutile. Et comment serais-je coupable de l'aider, lorsqu'elle appelle mon secours par les symptômes les moins équivoques? Ne la provoquons jamais; mais prêtons-lui la main dans l'occasion.'[57]

This is a very humble point of view, one that seems, on the face of it, to put Man and his mass of accumulated knowledge very much in its place. And yet there arises from this aspect of Diderot's thought, more than from any other, a genuine and almost poignant humanism. Despite Man's relative ignorance, and despite the leading role played by nature in general as a healer, it must never be forgotten that the point of medicine is to help human beings and to ameliorate the human condition. For Diderot this assistance meant attending not only to matters of physical welfare, but the moral as well. Because Diderot regarded people as purely material and conceived of 'the soul' as nothing more than that set of properties (consciousness, reason, passion, feeling) which arise from matter (organised in a particular fashion), he felt that the moral health of a patient was in fact quite inseparable from his physical health. Morals, as well as the more obvious physical ailments of the body, were therefore seen as the province of the doctor. Evidence of this view is widely disseminated throughout Diderot's work. In *Réfutation d'Helvétius* Diderot pronounces that it is the doctor that he consults about people's vices, virtues and dispositions.[58] *Eléments de physiologie* concludes by making a similar point; it emphasises that it is the doctors who hold the key to the nation's virtue.

Frustratingly these remarks are rather vague, if not cryptic. While it follows from the presumption that if the moral is conditioned by and dependent upon the physical then an activity that pertains to the physical well-being of a person should also pertain to their moral well-being, Diderot's comments on this subject tend to be quite cursory. He does, as shall be seen in due course, put forward a quite complex theory as to how the mind (of a purely material being) operates, but this level of sophistication is not matched by any statements regarding what doctors can or should do to encourage what might be termed 'moral health'. Diderot's view of the world would seem to permit doctors to usurp the role of priests, but while his canon provides a veritable mine of information regarding the subject of how medical men should go about the art of healing, there is precious little information about precisely how they are to act as the ultimate guardians and promoters of morality in society. For an issue which, one would think, is of paramount importance, Diderot's statements are surprisingly bereft of explanation and elaboration. Ultimately, a more thorough elucidation of such notions regarding the connection between medicine and the 'moral health' of society would

57. *Le Rêve de D'Alembert*, p.155.
58. *Réfutation d'Helvétius*, p.575.

have to await the rise of the ideologues at the very end of the eighteenth century.[59]

In any case, Diderot persisted in emphasising the importance of medicine as the best means by which to improve people's lives. In eminently practical fashion – which typified the climate of opinion that guided the Republic of Letters of which he was a part – Diderot's eyes remained firmly fixed on issues relevant to the progress of medicine and of human society. Of course, medicine could not progress without understanding, and it is here that it becomes essential to consider the discipline not as a practice but as a set of ideas. This is to make the leap from technology to science, from appreciating what medicine can do, to discussing the basis for why it works. It is here that further, more subtle and profound insight is gleaned as to why doctors were Diderot's favourite people with whom to speak.

Diderot's *Prospectus* to the *Encyclopédie* fits medicine into the 'Tree of Knowledge' under the umbrella of 'physique particulière'. As such, his description of the discipline's place follows the standard format: 'De l'histoire des *animaux*, prise par les sens, la réflexion a passé à la recherche de leur conservation, propagation, usage, organisation, etc., et a produit la science qu'on nomme *zoologie*; d'où sont émanés la *médecine*' (italics are Diderot's).[60] The *Prospectus* goes on to define medicine in accordance with the conception of Boerhaave, stating that it is occupied with:

> l'économie du corps humaine et *raisonne* son anatomie, d'où naît la *physiologie*; ou s'occupe de la manière de la garantir des maladies, et s'appelle *hygiène*; ou considère le corps malade, et traite des causes, des différences et des symptômes des maladies, et s'appelle *pathologie*; ou a pour objet les signes de la vie, de la santé et des maladies, leur diagnostic et prognostic, et prend le nom de *séméiotique*; ou enseigne l'art de guérir, et se subdivise en *diète*, *pharmacie*, et *chirurgie*, les trois branches de la *thérapeutique*.[61]

This five-fold division was quite standard, and presented a definition of the various medical disciplines and their interrelationships that had been entrenched since the seventeenth century.[62] Nevertheless these passages contain several points of interest, not the least of which is that the emanation of medicine from zoology reaffirms Man's place in the world as little more than just another animal. It is also worth noting that in following the divisions of Boerhaave, Diderot was paying homage to one of the most prominent names in eighteenth-century medicine. Most importantly however, the fact that medicine is subsumed under 'physique particulière' (along with regions of inquiry such as astronomy, geology

59. For a detailed study of the Ideologues (including discussion of Diderot's potential influence on their thought – a subject that bears further research) see Staum, *Cabanis*.

60. *Prospectus*, p.309.

61. *Prospectus*, p.309.

62. Brockliss and Jones, *Medical world of early modern France*, p.97.

and chemistry) denotes that it is but one branch of natural philosophy, one aspect of understanding the world of which Man is a part. This connection to the wider realm of natural philosophy is manifested in terms of both the methodology and content of medicine.

Methodologically Diderot viewed medicine in much the same way as any of the other sciences: as a fundamentally empirical discipline. While book learning and the systematisation of knowledge were important, Diderot felt that the best impetus to progress in understanding the human body was through observation and experiment. Ample testimony to this is found in *Plan d'une université*. Not only does Diderot devote a great deal of attention here to arguing that the training of doctors must be practically oriented, but he even remarks that 'J'ai quelquefois pensé que les charlatans qui habitent les faubourgs des grandes villes n'étaient pas si pernicieux qu'on le pensait. C'est l'empirisme qui a donné naissance à la médecine, et elle n'a de vrais progrès à attendre que l'empirisme.'[63] This rather radical view is strikingly reminiscent of Paracelsus, who admonished the world of scholarly medicine for summarily neglecting the practical wisdom, derived from years of experience, possessed by peasants, village healers and 'old wives'.[64]

The empirical bias of Diderot with regard to the advancement of medical knowledge therefore ran parallel to his biases pertaining to the actual treatment of patients. In terms of the latter, he stressed the importance of carefully observing the patient, assiduously surveying the circumstances, assembling as much factual knowledge as possible, and then proceeding to formulate a cautiously developed diagnosis. The acquisition of knowledge about the nature and operations of the human body in general followed much the same course. Information was to be gathered by the natural philosopher through a process of close observation and well-construed, precisely conducted experiments, and from the rich assembly of facts thus collected, hypotheses were to be induced with care and circumspection. Intuition of course also played a role in Diderot's methodology, but this was largely in terms of the formulation of hypotheses. Moreover, Diderot's notion of *pressentiment* was dependent upon the natural philosopher having immense and direct experience of working with nature; the intuition of his ideal man of genius was necessarily rooted in a close, symbiotic relationship with the object being examined. As has been demonstrated, the clinical method preferred by Diderot owed much to the influence and inspiration of Bordeu. By no coincidence, Bordeu was also for Diderot an exemplar of how medical research should be conducted. We have already seen how, in his 1751 research on glands, Bordeu employed careful anatomical study and an elegantly constructed experiment to demonstrate conclusively that glands

63. *Plan d'une université*, p.819-20.

64. Brockliss and Jones, *Medical world of early modern France*, p.120.

did not secrete fluid due to the contraction of muscles or the application of some other kind of mechanical pressure. This experiment is a perfect example of the kind of imagination, precision and sound construction of hypotheses that Diderot prescribed for experimental philosophy and the advancement of human knowledge. Thus Diderot's empirical attitude towards medicine as a science was in exactly the same vein as his attitude towards experimental physics as it has been delineated in chapter 2. The spirit of *L'Interprétation de la nature* was the very same spirit in which Man was to investigate matters of medicine as well as those of astronomy, botany or chemistry. Medicine was for Diderot subject to the same constraints and the same methods of structuring knowledge as any other branch of natural philosophical inquiry.

But the linkage between medicine and other branches of natural philosophy extended well beyond a shared methodology. According to Diderot's conception of the sciences there existed between medicine and the other disciplines several areas of substantive interdependency, shared content and issues of overlapping concern. In this crucial respect the realms that Diderot linked most explicitly with medicine were chemistry and natural history. This is made clear in *Plan d'une université*, which emphasises the interconnection between these three branches of learning on several occasions. The section on chemistry notes the indispensability of chemical knowledge to the doctor, and in the chapter dealing with the Faculty of Medicine, Diderot recommends that medical students be given preliminary lessons in both natural history and chemistry.[65] Significantly the connection goes deeper than concerns of mere pedagogy. Rouelle's lessons devoted tremendous time and effort to arguing that a greater knowledge of how bodies are constituted is essential to those that would take on the role of healer. It will be recalled, in fact, that part of Rouelle's very purpose was to displace botany with chemistry as the primary informant of medicine.[66]

The substantive nature of the connection between natural history and medicine was made manifest in Diderot's thought by his emphasis on the importance to health of environmental factors. If climate, geography – indeed all physical aspects of a given locality – were relevant to the diagnosis and treatment of a patient, then the doctor could not afford to ignore the riches provided by natural historical knowledge. Diderot's letter to Camper (*Observations et questions sur la variole*) makes this very apparent. At the centre of the inquiry are matters relevant not only to the manner in which the disease is actually spread, but questions about the

65. *Plan d'une université*, p.824-25.

66. Both subjects were of course taught in the medical faculties, at least to some degree. Rouelle's crusade was designed to shift the balance of power here dramatically in favour of chemistry. The fact that Rouelle's course was becoming increasingly popular with doctors in the middle of the eighteenth century would indicate that many physicians were inclined to agree with his point of view.

history of the malady and how it is affected by climate.[67] It was also crucial, according to Diderot, for medical practitioners to have some knowledge of the three kingdoms of matter, both in terms of how they are distinguished and how they are constituted. This is acknowledged in *Plan d'une université*, in which it is stated that natural history contributes to medicine knowledge of the three kingdoms of matter. While Diderot claimed on a number of occasions that in order to be a creditable natural philosopher one must also be a doctor, an unspoken assumption may have been that any doctor worth his salt must also be a naturalist.[68] According to Diderot, the disciplines may have differed in terms of scope and the precise direction of their attentions, but ultimately they could not do without one another.

The most obvious reason that medicine and natural history could not do without one another of course is that a key aspect of both sciences involved articulating a stance regarding the precise nature of life itself. In broaching this issue, it was not only natural history and chemistry that were implicated in a close, substantive relationship with medicine, but physics as well, since considering the nature of life invoked for Diderot and many of his peers nothing less than a full-fledged debate as to the nature of matter itself. Therefore, a key aspect of his world-view in terms of physics – the need to broaden the accepted definition of matter – played a role in determining the shape of his thoughts on medicine. Here this not only strikes at the heart of how all the disciplines of natural philosophy were intertwined, but at the core of what was one of the most lively and significant disagreements of the eighteenth-century French medical world, namely that which historians have referred to as the divide between iatromechanism and vitalism.[69]

For Diderot, as for his contemporaries, the manner in which they viewed and practised medical science depended to a large degree on how they regarded the body. It may be said that the view of the body depended upon two basic considerations: the nature of the matter of which the body was composed, and the mode of the body's operation. These questions are, of course, fundamentally linked, and eighteenth-century iatromechanists and vitalists offered explanations that seem in many respects diametrically opposed. It is worth noting, however, that iatromechanism and vitalism were not just different philosophical solutions to explain the nature of the human body, the nature of the matter

67. It has been suggested by James Riley that Robert Boyle's corpuscular theory of the atmosphere assisted in the development of the idea that air can be a vehicle for disease. See Riley, *Eighteenth-century campaign*, p.13.

68. *Réfutation d'Helvétius*, p.512.

69. For the appropriate secondary literature on this topic, see chapter 4 n.78, and also Rainer Schubert-Soldern, *Mechanism and vitalism: philosophical aspects of biology* (Notre Dame, IN, 1962), Kaitaro, *Diderot's holism*, and Roselyn Rey, *Naissance et développement du vitalisme en France de la deuxième moitié du 18e siècle à la fin du Premier Empire*, *SVEC* 381 (2000).

from which it was constituted and the means by which it originated, grew, developed and reproduced – they were also methodological ideals, or heuristic devices. These competing philosophies could be and were used as explanatory ideals or models from which doctors worked and upon which they based some of their most basic assumptions.

There are a number of ways to consider the distinction between iatromechanism and vitalism. At its simplest, the core belief of the iatromechanists was that organic bodies could be analysed, described and understood as nothing more than extremely complex machines in which the functioning of the whole could be accounted for in terms of the operation of its constituent parts. Man, in terms of his body at least, was nothing more than a robot comprised of organic matter, which was itself usually regarded as inert. Thus, bodily operations could be accounted for in terms of the Cartesian principles of matter and motion. Explanations of Man's higher faculties were usually another matter; they depended, according to most iatromechanical explanations, upon the soul, or some manner of spiritual substance distinct from the stuff of the body. It was hoped, certainly by some, that this conception of Man would provide the study of medicine with the same degree of rigour and certainty with which Newton had endowed physics. Mechanics and mechanical principles were therefore to be extended to every realm of human inquiry (with the exception, of course, of the soul, which remained autonomous) with equal efficacy and applicability. Iatromechanism therefore tended to rest upon at least three key pillars: first, that the matter which comprises the body is inert; second, that the operations of this matter, and the relations between bodily parts composed of it can be explained using the same principles as those applied to inorganic matter; third, that the whole is equal simply to the sum of its parts.

This is not to say that all iatromechanists adhered to precisely the same views. There was tremendous variation in terms of the way that iatromechanical theories of the human body could be construed. This point becomes readily apparent upon surveying the origins and development of iatromechanical thought in the seventeenth and eighteenth centuries. As has already been inferred, Descartes was in many ways the fountainhead of such thinking. Reducing everything in physics to the twin principles of matter and motion, Descartes allowed no room for the study of living forms as a distinct realm of phenomena; for him they could be subsumed under the same laws that governed all other types of activity. Animals, in Descartes's view, were hydraulic machines, the only thing redeeming Man from such a seemingly bleak appraisal being his immaterial soul.[70] The relationship between the brute matter which

70. These ideas were put forth in Descartes' *Traité de l'homme* (1648). For a translation and analysis of this important work, see René Descartes, *Treatise of Man*, translation and commentary by Thomas Steele Hall (Cambridge, MA, Harvard University Press, 1972).

composed human bodies and the immaterial stuff of his spirit would, of course, present Cartesian philosophers with tremendous problems – ones that were perhaps never adequately resolved. Nevertheless the Cartesian paradigm was significant, for it seemed to open up as never before the possibility of assembling a complete, coherent and materialistic science of the world and all things in it, including the human body.

The eighteenth century would erect great structures upon this groundwork, or a modified version of it. Within the realm of physiology, Descartes's 'bête machine' hypothesis would be extended to include Man himself, bringing everything that existed under an explanation that perceived the universe as operating in terms analogous to that of a man-made machine. This particular 'man-machine' thesis, of Julien Offray de la Mettrie, aroused tremendous ire in many mid-eighteenth-century circles and was heavily criticised not only on moral grounds (for its rejection of the idea of an immaterial soul) but on scientific grounds as well. Employing a method that has been described as predominantly 'speculative and deductive' La Mettrie posited that the human body was both a 'micro' and a 'meso' machine. That is, he believed not only that the operation of the whole organism could be reduced to those of its parts, but that the proper organisation of a given body converts the invisible vibrations or movements of the microscopic parts into visible movements, which were conceived to differ in different parts of the machine, and that therefore there existed a real and literal interdependence among moving parts.[71] Thus, in La Mettrie's view, organisms were explicable in terms of simple cause–effect relationships, depending upon the various interdependent material parts of the body.[72]

But not all approaches to medicine that might be described as mechanical were as strict and uncompromising as those of Descartes or La Mettrie. Robert Boyle, for instance, pondered whether or not the organisation of corpuscles (which was imagined to result in life) arises from a 'casual conflux of atoms' or from an 'extramaterial guide'. Uncertain as to where the truth of the matter lay, Boyle in any case adopted a mechanistic approach largely on the basis that immaterial principles or qualities should, when avoidable, not be admitted into the realm of explanation. For him, primary attention was to be paid to the minutest parts of the body as the principal agents responsible for the phenomena associated with living beings.[73] The great Dutch physiologist Herman Boerhaave maintained a similar position. He was less interested in the metaphysics of matter than in the observed relations between parts of the human body.[74] His approach may therefore be regarded as less absolute

71. Hall, *Ideas of life and matter*, vol.2, p.52. Although several decades old, this classic work remains one of the best general references on the history of physiology.

72. See chapter 4, p.157.

73. Hall, *Ideas of life and matter*, vol.1, p.283.

74. Haigh, *Xavier Bichat and medical theory*, p.24.

and more phenomenological than that of Descartes, but Boerhaave was probably the eighteenth century's most prominent and convincing member of the iatromechanical school. He helped to firmly entrench the idea that the human body is little more than a machine and that in time its operations may therefore become transparent to the human understanding in the same way that those of a clock can be grasped.[75]

By contrast vitalists asserted that the forces governing life were distinct from the phenomena of the inorganic world and therefore that the functioning of the human body could not be understood in stark, mechanical terms. As with iatromechanism, vitalism could be asserted on a number of different levels, or with varying degrees of commitment. Timo Kaitaro has posited two basic brands of vitalism, which he labels 'strong organicism' and 'weak organicism'. The former posits the actual existence of a 'vital principle' which directly accounts for the presence in organic matter of the capacities to feel, move and think, as well as for the correct operation of bodily parts. At the other end of the vitalist spectrum is 'weak organicism', which adopts an agnostic position regarding the exact nature of such forces, but nevertheless maintains that mechanical explanations alone are insufficient to account for the phenomena of life.[76]

This distinction becomes readily apparent upon surveying the views maintained by some of the eighteenth century's most notable proponents of vitalism. Georg Ernst Stahl (1660-1734), arguably the father of the doctrine, at least in its modern form, adopted an animistic perspective. Stahl posited the actual existence of a property or principle (akin to the soul) which could be viewed as having been placed in matter from without (perhaps by God), and which, being the prerogative of living things, distinguished them in an absolute sense from the inorganic world and therefore made the study of life something which rested on principles fundamentally different from those used to explain brute matter.[77] Others took what seems a more Newtonian tack, and while hesitating to specify the nature of the substance of life argued that we must suppose even if only hypothetically the existence of vital entities or dispositions. Boissier du Sauvages (1706-1767) for example maintained that just as mathematicians when solving problems had long utilised unknown quantities (for example *x* and *y*), so physiologists should have recourse to certain faculties even when ignorant of their essential nature and of the way they evoke their effects.[78] Thus, even though he refused to delineate the precise

75. This reading of Boerhaave meshes nicely with Jacques Roger's proposition that mechanism was, in the context of eighteenth century thought, a guarantee of the intelligibility of the world. See Roger, *Life sciences in eighteenth-century French thought*, p.359.

76. Kaitaro, *Diderot's holism*, p 28.

77. For more on Stahl and the particular nature of his animism, see Hall, *Ideas of life and matter*, vol.1, p.351-66.

78. Hall, *Ideas of life and matter*, vol.2, p.75. For more on the ideas and methodology of Sauvages consult Julian Martin, 'Sauvages' nosology: medical enlightenment in

nature of the 'vital principle', Sauvages very definitely believed in its actual existence and might therefore be termed an animist. Paul-Joseph Barthez (1743-1806), another famous Montpellier physician and exponent of vitalist doctrine, shared Sauvages' methodological circumspection (refusing to identify precisely what the life principle is) but he nevertheless maintained that some such principle really existed in matter and that the human being was a veritable composite of body, soul and vital principle. The commitment of Barthez went so far as to posit that this vital principle could actually be killed off by certain poisons without altering the body's corporeal structure.[79] This is a nice example, because it illustrates neatly what was at stake in the controversy over how the organism's capacity for feeling was produced: namely the question of whether life could be explained simply in terms of matter, or if its explanation could only happen with reference to special, super-added faculties.

Diderot's medical mentor Bordeu may be seen to have embraced a slightly more ambiguous position. He was a 'weak organicist' in that his objections to the mechanical paradigm were largely due to the experimentally demonstrated inadequacies of that theory. A 'strong organicism' comes across, however, in his notion of each bodily organ possessing a 'vital force' or mysterious principle of purposive movement that is responsible for its proper functioning. In this, his vitalism was fairly distinguished from that of his peers. While Barthez, for instance, maintained that the vital principle operated at even the most basic of levels (in modern parlance, we might say that of the cell), Bordeu was only really interested in discussing the manner in which it applied to more complex structures (such as organs). Also Bordeu tended to refrain from detailed metaphysical discussions regarding the exact nature of matter, and as such even his 'strong organicism' was less assertive than that of the animistic Boissier du Sauvages, and less thoroughly developed than that of Barthez. Indeed, Barthez occasionally criticised Bordeu for the degree to which the latter's theories still relied on elements of the mechanistic framework such as vessels, canals and friction.[80] The ground common to all of these doctrines, however, was the proposition that some manner of extra physico-chemical property is responsible for animating matter (either from within or without) and for actually performing the direction of functions that constitute 'being alive'.

Montpellier', in *The Medical enlightenment of the eighteenth century*, ed. Andrew Cunningham and Roger French (Cambridge, 1990), p.111-37.

79. Hall, *Ideas of life and matter*, vol.2, p.88.

80. Duchesneau, *La physiologie des Lumières: empirisme, modèles et théories*, p.385. For more on the discrepancies between Bordeu and Barthez, see Dominique Boury, 'Bordeu et Barthez: le malentendu vitaliste', in *La Lettre de la Maison Française d'Oxford*, no.14 (Oxford, 2001), p.71-84.

5. *Medicine*

Traditionally, scholars have branded Diderot an anti-mechanist and therefore aligned him most closely with the vitalist camp.[81] The reasons for this have been laid out in previous chapters. Chapters 2 and 3 have addressed Diderot's crusade to broaden the accepted definition of matter, his equation of matter with energy or force, and his emphasis on the capacity of matter to move itself. Chapter 4, of course, has discussed the development of Diderot's concept of *sensibilité universelle* – the proposition that all matter, even that which seems inert, is endowed with at least a trace amount of feeling (or the capacity to feel). This association of Diderot with a mode of thought which asserted that living matter is not inert but has a mysterious principle of purposive movement has led many to argue that for him (as for many other vitalists, most notably Barthez) medicine could become a discipline autonomous from the physical sciences.[82] Such a thesis has further led a number of historians to place Diderot in the front rank of those eighteenth-century thinkers responsible for the genuine emergence of the discipline that, at the turn of the century, Lamarck would label 'biology'.

This is, or would be, quite an accolade. But if the previous chapters have lent any credence to the view of Diderot as an anti-mechanist, they have equally demonstrated that this anti-mechanism must be carefully qualified, and that in the final analysis Diderot's philosophy is not at all easy to classify. Ample historigraphical evidence of this is the fact that recent scholarship has tended to emphasise the veritably mechanical aspects of Diderot's thought and work. The recent *Dictionary of eighteenth-century history* posits that 'In Newton's wake, mechanistic philosophies were most rigorously developed by French *philosophes*, like Diderot'.[83] In a recent posthumously published work Aram Vartanian takes a middle road, saying that at one point Diderot approached the study of the human body from a mechanistic point of view but that the concept of *sensibilité universelle* introduced to his physiology a significantly vitalistic dimension.[84]

In the same work however, it is astutely suggested by Vartanian that the categories of mechanism and anti-mechanism do not necessarily serve the historian well at all times and in all places. A plethora of ambiguities,

81. See, for instance, Yvon Belaval, 'Le Matérialisme de Diderot', in *Europäische Aufklärung: Herbert Dieckmann zum 60 Geburtstag*, ed. H. Friedrich and F. Schaml (Munich, 1967), p.9-21, and Ursula Winter, *Der materialismus bei Diderot* (Geneva, 1972). Jean Mayer's work also plays heavily upon this theme, as time and again it seeks to throw up instances of Diderot's resistance to the mechanical paradigm, or cites examples of ideas and thinkers that influenced or justified his supposedly vitalistic, anti-mechanistic thought.

82. Vartanian, 'From deist to atheist', p.48-49.

83. M. L. Benjamin, 'Mechanical Philosophy', in *A Dictionary of eighteenth-century history*, ed. J. Black and R. Porter (London, 2001), p.457.

84. Vartanian, *Science and humanism in the French Enlightenment*, p.116.

semantic subtleties and issues of context conspire to make the dichotomy between these two schools of thought, as well as the very definition of each, far too complex to be taken for granted. In the case of Diderot, perhaps better than anyone else of his century, it becomes apparent that if true understanding is to be had, the simplistic mechanism/anti-mechanism conundrum must be put aside, perhaps not completely, but at the very least into the background. Only by investigating thoroughly what Diderot has to say and the nature of the terms on which he says it, rather than through the employment of ready-made categories, can his subtle approach to the problem of life be made properly manifest.

One of the most notable manners in which Diderot's work belies straightforward interpretation regarding these issues is revealed in his language. Despite his development of the *sensibilité universelle* hypothesis, a notion that would seem, on the face of it, to be anathema to a mechanical interpretation of the world, the canon of Diderot's writings contain many statements about 'man the machine' or the 'mechanisms of the human body' which seem to either run counter to, or at least invite more careful consideration of traditional readings of him as sympathetic towards vitalisism. In *Réfutation d'Helvétius*, for example, Diderot remarks 'Les anatomistes, les médecins, les physiologistes vous le démontreront par un nombre infini de phénomènes: ouvrez leurs ouvrages, et vous verrez que ce ressort, quel qu'il soit, de toutes nos opérations intellectuelles souffre d'une manière presque miraculeuse de la moindre altération qui suivent dans le reste de la machine.'[85] Several lines later he asks Helvétius 'Vous persuaderez-vous aisément que dans *une machine telle que l'homme*, où tout est si étroitement lié, où tous les organes agissent et réagissent les uns sur les autres, une de ses parties, solide ou fluide, puisse être viciée impunément pour les autres?'[86] Similar allusions occur in *Lettre sur les sourds et muets*, where Diderot makes reference to thought itself operating in accordance with some mechanism.[87]

Are these statements to be taken at face value? Are they indicative of a conception of the human body as, essentially, a machine? There is the possibility of course that such remarks are merely rhetoric; that in employing words which refer to the body as a mechanism Diderot is simply dealing, perhaps somewhat thoughtlessly, in the linguistic currency of the period. This, however, would be an unsatisfying explanation. Moreover the prevalence of such language throughout Diderot's work and the potential importance of its ramifications mean that the contemplation the terminology that Diderot used should not be easily or lightly swept aside. The conception invoked by Diderot with the use of such language should be given rather more careful consideration, for there

85. *Réfutation d'Helvétius*, p.557.
86. *Réfutation d'Helvétius*, p.558.
87. *Lettre sur les sourds et muets*, p.526.

does exist the possibility that his mode of expression was indicative of a very real way in which a mechanistic vision of the human body pervaded his thought.

We might first consider here the fact that Diderot's espousal of *sensibilité universelle* was in fact highly qualified. As the previous chapter has made manifest, Diderot regarded the notion primarily as a useful hypothesis. *Réfutation d'Helvétius* and *Eléments de physiologie* make absolutely clear that the *sensibilité* theory is to be taken as just that; as theory and not as a statement of fact. Diderot is at pains in both of these works to distance himself from the fantasy that is *Le Rêve de D'Alembert*. He may have held that the idea of *sensibilité universelle* was imaginative and compelling, and that it solved a great number of philosophical problems at a stroke, but he never attempted to deceive either himself or his readers that it represented the truth, or that it even had any necessary ontological reality. What *sensibilité* ultimately represents within the framework of Diderot's world-view is not so much a cornerstone of his vision of reality, but an example of the highest degree of (potentially useful) speculation permissible within natural philosophy.

As an example of permissible speculation or hypothesisation within natural philosophy, the concept of *sensibilité universelle* was also for Diderot the prime example of how speculation was to be kept in check. The manner in which Diderot treated the idea reveals that for him, vitalistic explanations of the human body were really no explanations at all; to explain a function in vitalistic terms would have been to indulge in animism. Thus it is that in *Réfutation d'Helvétius* Diderot forcefully argued that *sensibilité* is not the cause of the phenomena of life, but a 'primitive condition'.[88] In response to Helvétius' contention that *sensibilité physique* is the cause of our actions, thoughts, passions and sociability, Diderot stated: 'Remarquez bien qu'il ne dit pas une condition primitive, essentielle, comme l'impénétrabilité l'est au mouvement, ce qui est incontestable; mais la cause, la cause unique, ce qui me semble presque aussi évidemment faux.'[89] This is a crucial argument, for it firmly established the parameters according to which the hypothesis could be applied and it ruled out the invocation of *sensibilité universelle* as an explanatory principle. Where traditional vitalists such as Barthez or even Bordeu insisted on some level that the 'vital principle' or 'vital force' dictates the proper functioning of bodily parts, Diderot was at pains to point out that *sensibilité* is only an underlying feature of matter – something that endows matter with the capacity to feel and to move, that is, something which accounts in only the most general of ways for the fact of 'animation'. Diderot was unequivocally opposed to the notion that such mysterious principles actually direct the functioning of the body.

88. *Réfutation d'Helvétius*, p.509.
89. *Réfutation d'Helvétius*, p.493.

According to Diderot, the specific causes and effects of bodily operations were clearly to be sought elsewhere; namely in the organisation of the human machine and the rapports which existed between its constituent parts. That this was the case is made manifest in *Réfutation d'Helvétius*, which places great emphasis on the relevance of organisation by arguing that while *sensibilité universelle* may, in theory, provide the root by which it is possible to believe that matter itself can 'live', the actual qualities of things (and it is in this that doctors are interested!) are determined by organisation – which can be studied in an experimental fashion.[90] The point is probably best demonstrated in *Eléments de physiologie*. This work made explicit the importance of analysing a body into the smallest possible constituent parts, and understanding it on the basis of how those affect one another and work in concert to account for the phenomena of human life. Everything is accounted for in terms of the material body. What truly concerns Diderot therefore is not so much the capacity for life, which is simply an act of redefining matter, but the elements of organisation by which the life of an organism is realised.

Of course, hanging on to his redefinition of matter separated Diderot from many iatromechanists, some of whom continued to regard the stuff of which the body was composed as completely inert. That Diderot sought to distance himself from this conception is apparent from such remarks as 'L'animal, suivant quelques auteurs, est une machine hydraulique. Que de sottises on peut dire d'apres cette unique supposition!'.[91] It should be noted however that for even the most die-hard iatromechanists the precise operation of the body was still a mystery; vast expanses of its nature were still regarded as largely uncharted territory. It was only the most optimistic and speculative of iatromechanists that presumed to have uncovered in a comprehensive way the true workings of the 'human machine'. In this sense iatromechanism was as much a heuristic device as it was an attempt to offer a complete and consistent account of the body. This significant distinction between a theory in theory and a theory in practice applied to vitalism as well, for as Jacques Roger noted, vitalists themselves never entirely dispensed with the notion that physics, chemistry and natural history had a role to play in explaining the body's operation.[92] Curious it is then that Roger himself once described Diderot's wavering between mechanism and vitalism as the 'énigme diderotienne'.[93] There is in fact no enigma, for there is no

90. *Réfutation d'Helvétius*, p.525.

91. *Eléments de physiologie*, Quintili edn, p.120. He based this statement, at least in part, on the propositions that there can be no perfectly hard bodies, nor any perfectly elastic bodies, and that the nature of fluid movement is still too little understood, as is the nature of the movement exhibited by sensible, animated, organised, living bodies.

92. Jacques Roger, 'The Living world', in *Ferment of Knowledge*, ed. Rousseau and Porter, p.277.

93. Aram Vartanian discusses Roger's remark in a book review (of the latter's *Les Sciences*

wavering. Diderot, like many doctors of his time, subscribed to significant portions of both schools.

This is not to embrace contradiction, but simply to put both iatromechanism and vitalism in perspective. We may say of Diderot's medicine that which may be said of his natural philosophy in general; that he holds seemingly divergent viewpoints not in opposition to one another, but as different levels from which to regard the same thing. Indeed Diderot's work and thought emerges as perhaps the best demonstration of the fact that the iatromechanical and vitalist points of view must be conceived not as hard and fast positions, but as points along a continuous spectrum of approaches to physiology. *Sensibilité universelle* is necessary in Diderot's thought, but mainly as a background element – it provides the metaphysical basis for making sense of life. Beyond this, its influence is weak, perhaps non-existent.

These issues come into better perspective when Diderot's thought is compared with that of La Mettrie and Maupertuis. Thomas Hall argues that while Diderot's physiology charts a middle route between the theories of La Mettrie and Maupertuis, it leans most heavily towards the conception of Maupertuis – whose vaguely animistic ideas Hall relates to Diderot's notion of *sensibilité*.[94] A truer appreciation of the situation however reveals that Diderot's ideas are actually very close to those of La Mettrie, and that the divergence between them ultimately rests on some very subtle points. La Mettrie asserted that motion and sensation are potential in matter, but was reticent about being drawn into discussion about whether or not they were essential qualities. In other words, he was careful not to hinge his view of life upon a proclamation of animating force that was seen to necessarily inhere in individual particles (thus, our discussion of living bodies must be confined to a consideration of their organisation). Diderot's position is simply that if these principles are to be seen as potential in matter (as indeed both Diderot and La Mettrie regarded them), then they must in some sense be conceived to inhere in matter at all times. It is as though Diderot adheres to this point of view because it seems to him the only logical, consistent position. Diderot's definition of *sensibilité universelle* therefore ultimately boils down to the almost tautological proposition that 'life is inherent in matter to the extent that matter has the capacity to live, or to participate in the arrangement of some body that exhibits the characteristics that are typically identifiable with "being alive" (such as sensitivity, motility etc.)'. Thus the distinction between Diderot and La Mettrie really rests on such a fine point that many might be tempted to consider their differences to be almost a matter of semantics, or at least of

de la vie dans la dix-huitème siècle) entitled 'The Problem of generation and the French Enlightenment', in *Diderot studies* 6 (1964), p.352.

94. Hall, *Ideas of life and matter*, vol.2, p.62.

methodological hair-splitting. In any case, in terms of understanding the human body, Diderot is immeasurably closer to the position of La Mettrie than to any of the nominal 'vitalists'. This fact has been underappreciated by most scholars, who have tended to overplay the vitalistic aspects of Diderot's theory, much to the detriment of not only those aspects of his thought that might be termed generally mechanical, but also his profound awareness of the theoretical, methodological and epistemological elements of the entire issue.[95]

Diderot's vision of the human organism is therefore better characterised (as Timo Kaitaro suggests) as being affiliated with 'holism' rather than with 'vitalism', for the properties that are emergent from organisation (consciousness, the unity of the self, for lack of a better term, 'the soul') are derived from and determined by the structure of the body, not some mysterious principle which animates or governs it.[96] If *sensibilité* achieves a higher manifestation in arrangements of greater complexity, this is not due to the impetus of *sensibilité* itself, but strictly down to the nature of the particular organisational structure. The elements referred to above may represent a higher, non-mechanical level of reality, but for Diderot they only do so as ephemeral, emergent properties, and are not regarded as being created and motivated by some supernatural, external or animistic force.[97] This is affirmed in *Réfutation d'Helvétius* when it is stated that the moral aspects of humanity may be able to constrain the physical, but they cannot direct it. The lines of influence work in exactly the opposite direction: it is the physical that dictates the moral. Mind, consciousness and 'the soul' are, for Diderot, entirely dependent upon the matter of which the body is constituted and the manner of its constitution. The 'holism' of Diderot therefore is simply realised as the principle by which the individual is unified – the whole does not direct the operation of the parts.

How then might mechanism, as an explanatory tool, be regarded within the framework of Diderot's thought? Quite simply, it can be

95. The relative proximity here of Diderot and La Mettrie has not gone entirely without notice. Ann Thomson, for instance, has acknowledged that their ideas were very similar, although she does maintain that Diderot hesitates somewhat between ascribing *sensibilité* to matter itself and regarding it as a product of organisation (in the manner of La Mettrie). See Ann Thomson, 'L'Unité matérielle de l'homme chez La Mettrie et Diderot', in *Colloque international Diderot* (Paris, 1985), p.61-68.

96. Kaitaro, *Diderot's holism*, p.23-24.

97. In maintaining the position that these higher faculties are emergent properties of the matter from which they arise, Diderot was assisted by the precedent set by John Locke, who asserted that an organised, sophisticated system of matter may acquire properties that a less complex system cannot have. Of course, Locke believed that even if there is a deductive, necessary relation between the nature of a substance and its qualities, the nature of those qualities is not necessarily fixed by the substance. By contrast, Diderot felt the relation between matter and mind to be one of supervenance (i.e. mind is dependant upon matter, but not vice versa). Regarding Locke's ideas on these issues, see John W. Yolton, *Thinking matter: materialism in eighteenth-century Britain* (Oxford, 1984), p.198-99.

posited as a higher (as in slightly removed from the concrete world of particular things), scientific level of explanation. On a practical, phenomenological level, the doctor does in fact treat the body as a machine. Parts are analysed, and its operations are explained in the same fashion as any of the other physical sciences: by accounting for the greatest number of phenomena with reference to the least number of causes (bearing in mind, again, the important point that *sensibilité universelle* is not a cause!). Mechanism, in Diderot's thought, may therefore be interpreted in much the same manner as his mathematics. Both are matters not of casting something out, but of qualifying it. Diderot's rebuttal of the overly optimistic 'hydraulic machine' hypothesis made it apparent that mechanism could not penetrate to and make assumptions about the most fundamental level of reality, just as he argued that it must be acknowledged that mathematics is not equal to reality but merely a matter of abstracting from reality. He nevertheless regarded both devices as very useful tools. Indeed, Diderot clearly felt that as higher levels from which to regard the terrain of reality, both were necessary to the advancement of human knowledge.

That Diderot felt this to be the case is made apparent when he states that it is to physics and chemistry that medicine must turn in order to learn more about the nature of life: 'J'invite tous les physiciens et tous les chimistes à rechercher ce que c'est la substance animale, sensible et vivante.'[98] The reference to chemistry here is of special significance, for while Diderot did shun iatromechanism, in its strict, literal, 'man is a hydraulic machine comprised of inert matter' guise, there is the possibility that he may also be described as an iatrochemist, or at least as adhering to a very iatrochemical point of view.[99] There is nothing in Diderot's physiology, in terms of the constituents of the human body or the way they operate, that is not in chemistry as he learned it from Rouelle. Even those aspects of medicine that might be described as vaguely 'vitalistic' in tone are all contained within chemistry, and perhaps even justified by it. That organic matter should be self-creative, that it should grow, nourish itself, undergo transformations, react with other bits of matter to create substances that display entirely novel qualities – all of this is within the parameters of chemistry as Diderot understood it. Moreover it could all be accounted for in a scientific manner and without references to external or 'occult' qualities. That which enabled a combination of substance 'A' and substance 'B' to form the compound 'C' was, qualitatively, not really different from *sensibilité universelle* – the 'primitive condition' (of matter)

98. *Réfutation d'Helvétius*, p.491-92.

99. Iatrochemistry was the doctrine that physiological phenomena could be explained with reference to chemical processes such as combination, fermentation, putrefaction or the activity of acids and alkalis. For more on this school of thought refer to Brockliss and Jones, *Medical world of early modern France*, p.144-45.

upon which life depended. Thus Diderot conceived of chemistry as not just a pharmaceutical aid to medicine, but as a metaphysical brother. Medicine, just like chemistry, depended for Diderot on the same process of simply broadening the definition of matter, while refusing to admit that this makes its observed operations inexplicable.

This then forms the starting point for Diderot's explanation of the human body and its operations: the potential for life itself must be viewed as a property of matter, but organisms themselves must still be understood in terms of their characteristic organisation. This careful balance is best demonstrated in Diderot's discussion of generation in *Eléments de physiologie*. Here, Diderot presents all of the major schools of thought regarding the problem of generation, giving a brief overview of animiculism, ovism, preformation and epigenesis.[100] His own position was somewhat ambiguous and he states in fact that a high degree of uncertainty surrounds issues of generation. If Diderot did tend to prefer one theory over the others however, it was most certainly epigenesis. That he leaned towards this view is demonstrated not only in *Eléments de physiologie*, but in *Le Rêve de D'Alembert* as well. Briefly stated, the epigenetic theory asserted that both father and mother contributed to the development of a child, and more importantly, that its original form was not a miniaturised version of its mature structure, but that the seed (originally a mass of undifferentiated matter) transformed itself to become a fully developed human being. In other words, a very simple structure developed into a complex structure over a period of time. Diderot illustrated the point by use of analogy, arguing that the feeble acorn does not at all resemble the mighty oak that it will become. In the case of both the oak and the human being, this transformation, though wondrous, is not taken to be an impenetrable mystery. As has been demonstrated in the previous chapter, Diderot presumed that the process of generation does follow a regular set of laws. He acknowledged that these laws may be virtually unknown, but was far from conceding that they are unknowable and, most importantly, he was absolutely convinced that such regularity must exist. In Diderot's view it was breakdowns in these underlying mechanisms of generation that accounted for gross 'monstrosities' or the severe departure of an individual from the 'norm' represented by the bulk of its species. Tantalisingly *Le Rêve de D'Alembert* also indulges in some speculation as to how such mutations might take place. Significantly this speculation conceives of the generation and subsequent development of beings in a profoundly physical way, in that the fibres of the embryo are taken to correspond to the organs of the fully developed adult. These fibres are, of course, not simply miniature versions of the adult organ, but over time they will assume a designated form in

100. The best overview of eighteenth-century theories of generation is undoubtedly provided by Roger in *Life sciences in eighteenth-century French thought*.

accordance with nature's laws. Mutations occur, according to the character Bordeu, when the fibres are damaged, distorted or receive some other form of interference. Thus, even in dealing with the perplexities of generation, Diderot attempted to hold off the classically sterile postulates of the preformationists, while simultaneously embracing a view that is pervaded with the seemingly mechanistic presumptions that the whole is essentially determined by its material parts, that inorganic matter is governed by physical laws and that it behaves in an ordered fashion that will eventually be more or less comprehensible to Man's understanding.

This argument is, of course, but one aspect of *Eléments de physiologie*, which is by far the richest source of information on Diderot's medical world-view. It is a voluminous text, one that duly reflects the mass and variety of medical knowledge that Diderot had acquired by late in life. *Eléments* is also a work that has received no small amount of praise. According to Diderot's literary executor J. A. Naigeon, the work was intended by its author to be a serious scientific text, and Jean Mayer noted that it is superior, in many respects, to most other such tracts that were available at the time.[101]

In assembling *Eléments de physiologie*, Diderot drew upon an exceptionally wide variety of sources, including all of the most significant medical minds of the century: Bordeu, Barthez, Bonnet, von Haller, Whytt, Boerhaave and Buffon are but a few of the most notable.[102] Although drawing both technical details and inspiration from many of these sources, Diderot did not, of course, treat them all in equal fashion. Generally speaking, one of the primary purposes of the book was to combat mind-body dualism, and therefore thinkers whose physiology tended to rely very heavily upon this theory, such as J. P. Marat, usually received fairly short shrift from Diderot.[103] But aside from matters of philosophical bias, which obviously played a crucial role in determining the shape of the book, the main goal of *Eléments* was to simply present the sum total of human knowledge that had been acquired up until the 1770s regarding the structure, function and interrelationships of various parts of the body. It may even be considered Diderot's second encyclopedia and, with good reason; it took at least eight years not to mention a lifetime of learning to assemble.

The first part of the tome is entitled 'Des Êtres' and, as the heading suggests, deals with the nature of life in general. The subject matter is more concerned with natural history and is covered, for the most part, in chapter 4. It includes a survey of the three kingdoms of matter, devoting special attention to the transitions between them. It is also in this section

101. See *Eléments de physiologie*, Mayer edn, p.lvi.

102. See above p.183-84.

103. On Marat's natural philosophical and physiological interests, see Clifford Conner, *Jean Paul Marat: scientist and revolutionary* (Amherst, NY, 1999).

that Diderot presents much of his proto-evolutionary theory. Significantly this section is permeated with chemistry, and deals at length with some of the most elemental forces which shape the world of living things; heat and fermentation, sensibility and irritability. Regarding the latter, Diderot sides with Whytt against Haller in maintaining that the two are intimately linked.[104] This portion of the *Eléments* gives at least a cursory survey of almost every basic aspect of life, from instinct to involuntary movement to the differentiation of animals by their organisation to the nature of life and death. The section culminates with a discussion of Man, which effectively serves as an introduction to the rest of the book as it quickly delineates Diderot's views on reason (the defining characteristic of the human species), thought and the soul.

The second part of *Eléments* deals with the 'Elements and Parts of the Human Body'. This very substantial portion of the work is the heart of Diderot's physiology; it describes and accounts for every organ and key operation of the body. It begins with a discussion of the most basic fibres that comprise human tissue. Diderot followed Boerhaave and Haller here to a very high degree, portraying fibres as the 'building blocks' of organic life.[105] Arguing that fibres are to physiology what the line is to mathematics,[106] Diderot saw fibres as exerting a continual effort towards contractility, and proclaimed that 'it is this tendency, this constant effort of the fibres, that is the chief means through which life is maintained'.[107] Thus in Diderot's physiology fibres can be taken to represent a manifestation of life of an order slightly higher than that of the *sensibilité* of the molecule. But where *sensibilité* was, as has been repeatedly stressed, simply a hypothesis relevant essentially as part of an almost metaphysical redefinition of matter, fibres represented life on a scale that could actually be observed and experimented upon.

From this basic element of the body, Diderot proceeded to offer a sweeping overview of the human machine, addressing each system and each organ in turn. The individual chapters tend to follow a fairly standard format, commencing with a physical description of the organ in question, followed by a discussion of its structure, function and, where possible, necessary connections to other organs or to the bodily machine as a whole. In this latter respect, Diderot's work leans heavily on that of Bordeu, and the idea that organs are semi-autonomous. Three distinct levels of life are identified: that of the molecule, that of the organ and that of the animal as a whole. Life, in other words, was conceived by Bordeu and Diderot as a 'multi-layered' phenomenon. Only in the case of the

104. See Hubert Steinke, 'Haller's concept of irritability and sensibility and its reception in France', in *La Lettre de la Maison Française d'Oxford*, no.14, p.37-69.

105. Hall, *Ideas of life and matter*, vol.2, p.61.

106. *Eléments de physiologie*, Quintili edn, p.156.

107. Hall, *Ideas of life and matter*, vol.2, p.61-62.

lone molecule, Diderot tells his reader, is life never completely extinguished. On a higher level, these molecules are amassed to form fibres and eventually organs, which act to a certain degree as complete systems in their own right. Each organ experiences its own states of health and sickness, and its 'life' can be defined in large part according to the function it performs. The life of the whole being is, in turn, a matter of the balanced and harmonious activity of these organs.

Interestingly Diderot marvels at how few organs are actually essential to human life. In performing his survey, however, he identifies three organs that overshadow the others and form something of a triumvirate that governs the body. These 'great animals' or 'great centres of movement' are the stomach, the heart and the brain.[108] In this, Diderot was following in the footsteps of Bordeu, although he departed from his mentor in also emphasising the significance of the diaphragm.[109] This latter organ, described as a muscular membrane, is seen as the centre of feeling and of passion, the seat of our pleasures and pains.[110] As such the diaphragm is, in Diderot's opinion, that which makes men compassionate and moral.[111] Diderot's remarks here are generally vague and there is as little elucidation regarding the precise nature of the connection between the diaphragm and 'feeling' as there is in terms of the proposition that doctors can tend to the moral well-being of society. Although he does note that there exists a sympathy between the diaphragm and the brain, no elaboration is offered as to how this might work.[112] Diderot's primary point here seems simply to be that people often feel their emotions in or via the diaphragm. The experience of joy or of sorrow, for instance, is often referred to this organ; sensations occurring there are part of the experience of emotion as a whole. The connection would appear to be as poetic as it is scientific or philosophical.

The organ that is by far and away the most important to Diderot however is the brain, the organ of thought, to which the entire third section of *Eléments* is devoted. Part of the reason for the profound significance of the brain according to Diderot's system stems from the fact that his medical and natural historical world-view tends to make the distinction between the individual and the whole very ambiguous. This has been witnessed on many counts: the essentially unified and interdependent nature of the universe as outlined in Diderot's cosmology, the delicate interplay between an organism and its environment, and of course, the complex and multi-layered structure of the organism itself, in which even the constituent organs and solitary molecules can be con-

108. *Eléments de physiologie*, Quintili edn, p.187.
109. Haigh, *Xavier Bichat and medical theory*, p.36.
110. *Eléments de physiologie*, Quintili edn, p.211.
111. *Eléments de physiologie*, Quintili edn, p.211.
112. *Eléments de physiologie*, Quintili edn, p.211.

sidered as possessing a certain kind of autonomous existence. The entire concept of individuality thus becomes somewhat arbitrary. We can, in the abstract, draw concentric circles within circles of identity almost ad infinitum. Where does any particular individual begin, and where does it end? To a certain extent, in the universe according to Diderot, it does not. Everything is necessarily linked to everything else.

Of course, such an abstract, poetic, metaphysical view was just one aspect of Diderot's thought on the matter. In a more concrete sense, he did seek to identify certain stages of individuality. A hint of the manner in which he did this is presented in *Le Rêve de D'Alembert*, which posits that molecules can be either contiguous or continuous with one another. Contiguous molecules exist merely side by side, their motions can and do affect one another, but they are not united in any sense such as to constitute a whole on their own. Continuous molecules however are more intimately linked; they come together much as two globules of mercury meld to form a single, larger drop. It is in this state of continuity that organs exist, with all of their constituent molecules acting in some kind of concert towards a single goal or at least as part of a given, semi-closed system. On an even higher level, it is the capacity of molecules to become continuous with one another that unites the organism as a whole. As Diderot explained in *Le Rêve de D'Alembert*:

> Voulez-vous transformer la grappe d'abeilles en un seul et unique animal? Amollissez les pattes par lesquelles elles se tiennent; de contiguës qu'elles étaient, rendez-les continues. Entre ce nouvel état de la grappe et le précédent, il y a certainment une différence marquée; et quelle peut être cette différence, sinon qu'à présent c'est un tout, un animal, un, et qu'auparavant, ce n'etait qu'un assemblage d'animaux [...] Tous nos organes [...] ne sont que des animaux distincts que la loi de continuité tient dans une sympathie, une unité, une identité générale.[113]

But even this conception falls short of offering a precise explanation as to exactly what makes an individual human being an entity separate from all that surrounds him. Diderot's distinction between continuity and contiguity is, after all, fairly vague, and amounts to little more than a metaphysical description. This may work in *Le Rêve de D'Alembert*, but for the purposes of *Eléments de physiologie* it is insufficient. It is here that the primacy of the brain in defining the nature of human life comes into play, for it is ultimately on account of this organ that individuals are able to acquire their identity. Memory, one of the most important functions of the brain described by Diderot, is what truly gives an individual a sense of their own separateness. In being able to link past moments with the present, the life of a person assumes the quality and continuity of a story; it constitutes a certain kind of whole unto itself.

113. *Le Rêve de D'Alembert*, p.83-84.

The brain was also for Diderot the centre of a great network. It is, to use Diderot's analogy (itself borrowed from Bordeu), like a spider to its web. The web, of course, was meant to represent the nerves. Described as a substance 'médullaire du cerveau', exposed to the action of external bodies, nerves were seen by Diderot as the principles of sensitivity and action in the human body, the organs of movement and servants of the brain.[114] In other words, nerves were presumed to act as conductors of some sort for both transmitting sensations to the brain and for inciting activity in the muscles; indeed Diderot argued that without the intervention of nerves, there would be no sensation.[115] Regarding the nature of nerves themselves, he followed in the footsteps of Bordeu by asserting that they are a quite unique element of the animal body. Despite their responsibility for transmitting feeling, nerves, unlike muscles, were not regarded as irritable in themselves (that is, they are not visibly agitated 'under the scalpel' as muscles are).[116] Thus they acquire here a rather ambiguous status – not quite like any other element of the body, and yet responsible for so much of what actually happens within it. Crucially the means by which nerves act was still very much a mystery for Diderot, as it was for his contemporaries. Certainly, he knew which theories he did not like. Using the unfortunate Marat as his straw man (which happened quite frequently in *Eléments de physiologie*!), Diderot criticised the notion that nervous operations are carried out by means of a fluid (with part of the brain acting as the filter and another part as a reservoir).[117] Instead he seemed to lean towards the school of thought which held that the nerves act by vibration.[118] Part of the reason for this preference was his inability to see how, given what was known about the constitution of nerves, they could possibly be viewed as fluid-bearing vessels.[119] In any case Diderot's emphasis on the importance of the nervous network is another feature of his physiological view that provides a certain unity to the organic being. The brain, centre of a vast, purely corporeal web, is able to respond to and act upon all of the other parts of the body.

But this coordinating function was just one of the crucial functions performed by the brain. Perhaps the most important function is reason, which Diderot denoted as being the aspect of human existence that

114. *Eléments de physiologie*, Quintili edn, p.176.

115. *Eléments de physiologie*, Quintili edn, p.178. Jean Mayer notes that in holding this opinion, Diderot was opposing the view of Barthez (who, in his thoroughgoing vitalism, considered every part of the body to be directly capable of sensation). See *Eléments de physiologie*, Mayer edn, p.92.

116. *Eléments de physiologie*, Quintili edn, p.175.

117. *Eléments de physiologie*, Quintili edn, p.181.

118. For an interesting and informative discussion regarding the development of such ideas, beginning with the speculations of Newton, see Yolton, *Thinking matter*, especially the section entitled 'The Aether and vibrations', p.177-85.

119. *Eléments de physiologie*, Quintili edn, p.182.

separates Man from animals. Each animal in nature tends to be governed by a particular instinct, and reason, Diderot tells us, is the proper instinct of Man. It is due to this capacity that understanding and knowledge can be acquired, and moreover that they can be developed and enhanced in such a way that humanity is able to improve itself.

Reason consists of the ability to link disparate ideas together and to form judgements. In true sensationalist style, it depends upon the evidence of the senses for its raw material. According to Diderot, there can be no thought without sensation. This, of course, should by now be an obvious point. Following very much in the tradition of Locke and Condillac, Diderot asserted that there are no innate ideas, but that all thought is essentially a matter of processing (separating simple ideas from complex ones, linking various ideas together, performing judgements upon said ideas etc.) the information received from the senses. It is unsurprising therefore that Diderot's section on the brain actually begins with a description of the senses. Only after this does it proceed to describe how that which is sensed is dealt with by the understanding, and thereafter used by memory, imagination, the will and the passions.

There are a number of interesting points that arise from Diderot's discussion of the senses. The first is that things are not always presented to the perceiving subject in a simple way, but are often part of complex arrangements. The senses are not held by Diderot to act independently of one another, but in concert. There is a reciprocal and continually varying interrelationship between sight, hearing, touch, smell and taste. More significantly perhaps, these sensory powers exist in the human species in a relative state of balance. Diderot argues that animals tend to be dominated by one sense or another, and that this dominance determines the character of the animal, making it, in a very real way, a slave to a particular sense. Dogs, for example, are conceived by Diderot as little more than noses; they are creatures governed by their extremely good sense of smell. Humans, on the other hand, have no predominant sense. It is on account of this that the brain is able to take over the governance of the body, the understanding mediating carefully between the varied bits of information it receives from the outside world.

Eléments de physiologie emerges as a *pièce de résistance* for Diderot. It is a powerful and coherent statement of sensationalism and materialism. Everything about the body, and most importantly even about thought, is explained in purely material terms. Diderot therefore achieves a complete survey of human life, from the mysteries of the most elemental level, to Man's higher cognitive faculties; all are subsumed within one grand and purely materialistic world-view. This survey possesses a two-fold nature. In one respect it is a story of evolution, of the development of Man and society from the bare rudiments of matter itself. In another respect, it simply serves as a credible account of how even the highest and most abstract forms of thought can and do arise in a purely material universe. To fully appreciate how deeply this conception pervaded Diderot's thought, we must look beyond *Eléments de physiologie*, for this story, as it were, is found throughout much of his work and in many guises. *L'Interprétation de la nature*, for instance, surmises that life originated at the molecular level, and that through a process of rising from the simple to the complex, bodies were formed and eventually became capable of feeling, sensation, thought, judgement and finally civilisation. More detail with respect to the actual formation of ideas, and languages in particular, is provided by *Lettre sur les sourds et muets*:

> Les objets sensibles ont les premiers frappé les sens, et ceux qui réunissaient plusieurs qualités sensibles à la fois ont été les premiers nommés: ce sont les différents individus qui composent cet univers. On a ensuite distingué les qualités sensibles les unes des autres; on leur a donné des noms: ce sont la plupart des adjectifs. Enfin, abstraction faite de ces qualités sensibles, on a trouvé ou cru trouver quelque chose de commun dans tous ces individus, comme l'impénétrabilité, l'étendue, la couleur, la figure, etc., et l'on a formé les noms métaphysiques et généraux, et presque tous les substantifs. Peu à peu, on s'est accoutumé à croire que ces noms représentaient des êtres réels.[120]

It may be observed however that Diderot's views on these matters became more sophisticated and technical over time. Only with *Eléments de physiologie* does his sensationalist theory take on a completely scientific guise; thoroughly addressing all available knowledge regarding each sense and each function of the brain, and suggesting the manner by which the brain may receive information and interact with the rest of the body.

Of course, this nascent psychology carries with it some very profound philosophical implications. Not the least of these is the matter of determinism. By maintaining a strictly materialistic view of human physiology, in which the emergent properties encapsulated by the word 'soul' are simply derived from and unable to affect the body from which they arise, Diderot opened himself to the charge that free will cannot exist. The manner in which he treated this charge was somewhat ambiguous. Perhaps on account of his innate humanism he was unable or unwilling to adopt what would anachronistically be termed a behaviourist point of view. Nevertheless Diderot was ultimately unable to escape the necessary consequences of his materialism. As he put it: 'Après cela, je ne vous dirai de la liberté qu'un mot; c'est que la dernière de nos actions est l'effet nécessaire d'une cause une, nous, très compliquée, mais une.'[121] While not discussing at length a completely reductionist determinism, Diderot maintained that everything happens for a reason, that every effect has a cause and that every decision can be traced back to a purely corporeal explanation of thought. In this respect, *Jacques le fataliste* is the literary

120. *Lettre sur les sourds et muets*, p.522.
121. *Le Rêve de D'Alembert*, p.142.

equivalent of *Eléments de physiologie*; both works portray Man as fundamentally unable to transcend his material composition.[122]

Connected to this is the issue of mental health. Because, in Diderot's view, that which happens to the body affects the brain and therefore the manner in which we think, the state of our cognitive faculties and even our ability to reason are directly influenced by environment and by the current state of the body's operation. Diderot observed, for example, how the presence of a fever can impede or at least alter the capacity of a person to think properly. To study and maintain the health of the body was also therefore to study and maintain the health of the mind. As *Réfutation d'Helvétuis* makes clear however, differences in basic organisation were also responsible for varying modes of thought. While acknowledging the seminal role played by education and environment in the development of human character and ability, Diderot also stressed that to a certain extent no amount of education could ever overcome basic differences in the structure of the brain.[123] This led Diderot to the elitist conclusion that some of the highest truths (for instance those contained in Newton's *Principia mathematica*) could never be equally appreciated or even grasped by all, no matter what. Thus the importance of medicine in Diderot's natural philosophy was intensified on account of its ability both to understand better (and hopefully to control) the impact of the world upon the human organism, and to grasp better the nature of the differences in organisation which led directly to differences in things like cognitive capacity.

Medicine also raised for Diderot the question of whether or not there existed a universal human nature. Presumably, because Diderot placed such emphasis on organisation, scope was created to assert that there was a nature common to all human beings. The primary question of course, was trying to determine precisely what this nature was. The answer that emerges from Diderot's own work is very general. He held out reason as being the primary distinguishing characteristic of the human species. As we have seen, Diderot referred to reason as the 'proper instinct of Man'. Attached to this was the presumption that with reason came the possibility of perfectibility, although the latter concept was not actually defined by Diderot. At the very least, it may be understood that Man's reason allowed him to make ongoing improvements in his condition. Perhaps Diderot's most interesting statement regarding human nature

122. Of course *Jacques le fataliste* presents more than one point of view. As the title suggests however, the deterministic attitude of Jacques is one of the primary themes around which the work revolves. For analysis of this eighteenth-century classic, see John Loy, *Diderot's determined fatalist: a critical appreciation of Jacques le fataliste* (New York, 1950) and Anne Chevalier, *Jacques le fataliste* (Paris, 1993).

123. Thus the essence of Diderot's attack on Helvétius in this work is not on account of the latter's materialism (Diderot, of course, was fully supportive of this), but due to a perceived oversimplicity.

however comes in *Réfutation d'Helvétius*. Here, in response to Descartes's 'cogito ergo sum', Diderot proclaims that he prefers the solution of Hobbes, which is translated as 'je sens, je pense, je juge, donc un portion de matière organisée comme moi peut sentir, penser et juger'.[124] This succinct argument encapsulates much of Diderot's world-view. To think of human nature, indeed, humanity in general, is for Diderot to do nothing more than to make the empirical statement that beings organised in a certain similar way possess the same kind of characteristics.

But the most important aspect of medicine for Diderot's world-view lies in its epistemological implications. It has by this point been stressed repeatedly that according to Diderot the physical provided the grounding for how we think. The entire body of human knowledge, from the most fundamental sensation to the most intricate abstraction, relied solely on matter and the manner in which it was arranged. Medicine therefore provided the basis for coming to terms with the entire structure upon which natural philosophical knowledge of the world was built. Paradoxically medicine itself was a branch of natural philosophy, so medical knowledge itself was dependent upon and conditioned by an understanding of how the human cognitive faculties actually operate. For Diderot then, medical science was at once the pinnacle of natural philosophical knowledge and its very foundation. This, more than anything else, was the basis for Diderot's relativistic world-view, for his acute awareness that anything that bears the label of 'knowledge' must be qualified as 'human knowledge'; itself transient, limited, and entirely reliant upon the physical structure and nature of the human organism.[125]

Such a conception may have devalued the entire notion of knowledge for many of Diderot's contemporaries, but for Diderot, it was the only type of knowledge possible; it was simply considered to be the lot of humankind and the only terms upon which the universe could be confronted. But despite this objectification of Man and the subsequent subjectification of Man's knowledge, there is a respect in which Diderot's philosophy elevates humanity. Since the universe was ultimately regarded by Diderot from a holistic point of view, and since an individual could be defined at almost any point, the most significant effect of the brain's primacy according to Diderot's vision was to make the human individual the most important type of individual of all. It was only the human individual that could, through the faculty of understanding, gain a true appreciation of its differentiation from the matter around it, and further, attempt to comprehend the nature of that matter. It was the human being alone that could assemble any kind of picture as to the workings of

124. *Réfutation d'Helvétius*, p.491.

125. This notion is emphasised in *Lettre sur les sourds et muets*, in which Diderot states that only God could have a true perspective on the relationship between observer and observed. See *Lettre sur les sourds et muets*, p.574.

the universe of which it was a part. In Diderot's godless universe, this made the only kind of understanding, creativity or wisdom that existed indelibly human.

Medicine therefore played a plethora of roles within Diderot's natural philosophy. To Diderot's own statement that doctors were 'his favourite people with whom to speak', it may be added that they were likely the people with whom he most needed to speak as well. Without a thorough grasp of medicine, Diderot's natural philosophical framework would have been profoundly incomplete. Apart from its obvious practical applications and social value, medicine provided for Diderot a deeper understanding of organic bodies, and the manner in which they were constituted. More specifically, it is with reference to medicine that Diderot's stance regarding the mechanist/vitalist controversy comes into focus. Diderot's physiology managed to transcend the dichotomy between iatromechanism and vitalism much as the scientific methodology outlined in *L'Interprétation de la nature* transcended the oversimplistic opposition of rationalism and empiricism.[126] By introducing the hypothesis that *sensibilité* is universal, Diderot not only eliminated any absolute distinction between organic and inorganic matter, but effectively sidestepped the entire debate between iatromechanism and vitalism. He simply wound up positing (hypothetically, and as a primitive condition rather than as a cause) another force of nature which is intrinsic to matter (alongside gravity, electricity, magnetism etc.) and which amounts to making the statement that matter has the capacity for life, or better, that it is animated in such a way that given a particular organisation it will constitute a living being. In taking no more than this as his starting point for analysis of the 'human machine', and thereby attempting to offer a very mechanistic explanation of the manner in which that machine operates, Diderot was approaching the discipline of medicine from a perspective far less unorthodox than is generally considered.[127]

This approach, of course, is only truly appreciable in his later texts, most notably *Réfutation d'Helvétius* and *Eléments de physiologie*. Diderot's earlier works are far more vague and metaphysical. This rise in the confidence, sophistication and technicality of Diderot's expression illustrates not only that he was continually acquiring and applying new medical knowledge, but that his medical thought was developed in conjunction with all of the other scientific disciplines that captured his interest. His ideas evolved over time and as a result of fruitful interplay

126. See chapter 2.

127. Such a conclusion is also affirmed by the earlier part of this chapter, in which it was demonstrated that many of Diderot's ideas about the nature of medical practice and the role of the physician in society were actually quite commonplace. Although usually staying quite close to the ideas of Bordeu, these ideas themselves were more often than not part of the general Hippocratic tradition adhered to by many eighteenth-century doctors.

between all aspects of natural philosophy. It is only on account of having appreciated and elucidated this interconnectedness, and having viewed Diderot's 'biological' ideas from the perspective of his natural philosophy as a whole, that we have been able to situate him here regarding the contentious mechanism/vitalism divide.

Medicine therefore evolved in Diderot's thought from being a matter of mere curiosity to a consuming interest which spurred him to the assembly of an intricate physiological model of the human body, and a complex, sensationalist psychology which formed the basis of his epistemology, and in so doing made greater sense of all the other disciplines. Medicine, while perhaps less independent than any of the other branches of learning treated by Diderot (in the sense of being more reliant for its content upon the other disciplines), turns back upon them and confirms Diderot's own interpretation of the natural philosophical project as a whole. In this way Diderot's unified, natural philosophical outlook is brought full circle from questioning the terms upon which Man abstracts from reality in the formation of his ideas, to the elucidation of a methodology for regarding and dealing with nature, to a particular description of the human machine which ultimately makes sense of how and why we must think and assemble systems of knowledge as we do.

6. Conclusion

AMONG an earlier generation of Diderot scholars there emerged a debate regarding the exact place of science within the context of Diderot's life and work. On one side of this debate lay Lester Crocker, who maintained that Diderot's natural philosophical thought was dependant upon and superseded by his atheism – his science, in other words, was but a prop to his irreligion, a way of filling certain gaps present in a universe already presumed to be godless.[1] By contrast, Arthur Wilson, Philip Furbank and others asserted that Diderot's natural philosophy was an altogether more central aspect of the *philosophe*'s world-view, and that it could be viewed as independent of rather than as an adjunct to his atheism.[2] From discussion of the evidence in the preceding pages it should now be clear that the latter stance is probably much closer to the mark. At the very least, it may be observed that rather than Diderot's atheism being primary to his natural philosophy or vice versa, both of these crucial aspects of his philosophy were motivated by a third, perhaps deeper element: his humanism. What else, if not this, is revealed by his determined atheism? And what, if not a desire to eliminate supernatural concepts from the realm of philosophical enquiry and to place Man at the centre of an inherently meaningless universe, is one of the driving forces behind Diderot's science?

Of course, the thesis that both Diderot's atheism and his science were in some sense reducible to the more fundamental attribute of his humanism is an idea that merits greater attention than can be afforded in the pages that remain here. What matters most in the context of the present study is not simply that Crocker's position seems untenable, but also that in positing the above 'humanist thesis' we would have pulled off a very Diderotian move. It has been observed on several occasions in the preceding chapters that a tendency continually manifested in Diderot's philosophical thought was to broach dichotomies or seemingly intractable problems not so much by favouring one side of the argument over the other, or even by seeking to synthesise them, but by achieving some manner of transcendence regarding the whole debate. This has been witnessed, for example, with regard to mathematics versus experimental philosophy, the Cartesianism/Newtonianism conflagration, and of course Diderot's effective side-stepping of the mechanism–vitalism debate.

Given this tendency to approach the discussion of key concepts and issues by adjusting the very framework in which they are discussed, we

1. Crocker, *Diderot: the embattled philosopher*, p.317.
2. See Wilson, *Diderot*, and Philip Furbank, *Diderot* (London, 1993).

may well see Diderot as playing around the edges of the mid- to late-eighteenth-century natural philosophical paradigm without quite shattering it or sketching out an entirely new one. To borrow the terminology of the late Jacques Roger, we might view Diderot as probing the edges of but not bursting through the boundaries of the existing 'mentalité'.[3] Such a notion plays nicely into the relatively popular image of Diderot as a maverick, as a courageous thinker on the periphery of scientific thought. Indeed, this is how Diderot has traditionally been read with regard to almost every sphere of his interest, the most recent example being Jonathan Israel's voluminous *Radical Enlightenment*, which portrays figures like Diderot and La Mettrie as the culmination of a century-long tradition of subversive and unorthodox thought. But is this image entirely accurate, and does it truly assist the scholar in appreciating the import and nature of Diderot's natural philosophy?

One counter to this view of Diderot 'the radical' is that he actually interacted a great deal with mainstream science. He was, it will be recalled, a regular reader of the memoirs of the Académie des sciences. Many of those scientists who were closest to him, and from whom he learned much, were or became respected members of the French scientific establishment. Rouelle, for instance, was responsible for educating almost the entire generation of late-eighteenth-century chemists, including the illustrious Lavoisier. Buffon, though he did eventually scorn the Académie, assumed the leadership of a significant portion of French natural philosophical life in his role as administrator of the Jardin du Roi, and achieved a very notable degree of fame along the way as his *Histoire naturelle* went on to become virtually synonymous with the discipline of natural history. Bordeu, while hailing from the factory of generally unorthodox medical thought known as Montpellier, did attain a prominent position in Paris and was a well-respected, top-rank physician in his day. Beyond this core of mentors, Diderot had contact with and was influenced by an extremely wide range of mechanicians, physicists, experimentalists and medical professionals whose views spanned the spectrum from relatively radical to consummately mainstream.

That Diderot was not merely a dark horse on the periphery of eighteenth-century scientific thought but that he engaged frequently and fruitfully with the scientific establishment is also suggested by the trajectory of his scientific career from the 1740s to the 1770s. He took natural philosophy seriously, and his assiduity only seems to have grown with the years. Many ideas that were grasped and expressed only vaguely early on were eventually developed into firmer statements: more technical, more reliant upon what was regarded as proper scientific evidence, and more integrated as part of a broadly based, carefully developed natural philo-

3. Roger, *Life sciences in eighteenth-century French thought*, p.xxxii-xxxvii.

sophical world-view. Thus in mathematics Diderot's stance proceeded from that of straightforward love of the subject to a philosophical appreciation of the nature, meaning and limits of abstraction. With regard to physics, Diderot went from simply entertaining a number of questions regarding the nature of matter, to articulating a complex and deeply considered theory pertaining to how all matter should be regarded, as well as delineating and demonstrating an intricate scientific methodology. Despite suffering early in his career from a complete lack of knowledge regarding chemistry, Diderot eventually attained a very thorough grounding in the discipline and was able to make incisive use of such learning to enlighten his other philosophical pursuits. Diderot's natural history consisted in moving from the almost poetic expression of general cosmological ideas to articulating some of the eighteenth century's most intriguing views regarding the nature of the life, the universe and Man's place in it. Medicine was intertwined with the development of all these disciplines, and it evolved in Diderot's thought from being a matter of curiosity to a consuming interest which spurred him to the assembly of an intricate physiological model of the human body and a complex, sensationalist psychology which formed the basis his epistemology, and in so doing made greater sense of all the other disciplines.

More significantly, the content of Diderot's thought within each of these disciplines need not be seen as entirely unorthodox and in many respects might be viewed as rather conservative. Consider for example his passionate diatribe against the behaviouristic egalitarianism of Helvétius. Despite being an avowed materialist, Diderot's stance in this work, along with notions developed much earlier such as that of the natural philosophical 'genius', lend themselves to a reading of the *philosophe* by which he appears veritably elitist. The political, social and cultural overtones of much of Diderot's science are not what may easily be described as revolutionary.

All of this leads nicely into something else that should be noted about Diderot's natural philosophy, namely that his aforementioned tendency to transcend need not be seen as an inherently radical or even paradigm-challenging activity. A better way of looking at it is to see Diderot as staying comfortably within the bounds of mainstream science, while merely suggesting some crucial redefinitions. The most notable instance of this is his redefinition of matter, discussed at length in chapters 2 and 3. One might see him as simply re-prioritising certain items already within the scope of the primary currents of the day. Methodological circumspection comes before speculation, such that when investigating the confluence of matter and force we should not presume to know exactly how matter is constituted, but rather deal simply with observed effects. In a similar vein, discussion of 'general effects' supplants discussion of 'causes'. In addition to such key points, Diderot was also creating scope for the elaboration of a method involving creative, intuitive hypothesising.

Thus rather than trying to orient Diderot's natural philosophical career in terms of the apparatus provided by Thomas Kuhn's 'paradigm shifts', a more useful model may be Imre Lakatos' 'scientific research programme'.[4] Viewed in this light, the 'hard core' (the set of non-negotiable, foundational concepts and propositions) of Diderot's programme could be seen as constituted by his insistence upon the perpetual dynamism, timelessness and neutrality of the universe, the self-creative nature of matter, and an epistemology whereby Man has no direct access to the essential truths of nature, but only the ability to approximate it and build models on the basis of sensation, to construct knowledge that may be useful but is nevertheless asymptotic to reality. The 'positive heuristic' of such a research programme would then be comprised of manifold hypotheses, some expressed with greater confidence and certainty than others. Ideas regarding how bodies operate on a chemical level and the notion of evolution, for example, would be fairly close to the centre of the programme, while slightly further out would be concepts such as *sensibilité universelle*.

This idea is appealing because it meshes nicely with Diderot's vision regarding the linkage and purpose of each scientific discipline. As we have seen, each realm of natural philosophy served for Diderot a different purpose yet was still part of an overarching attempt to impose some kind of order on reality. As such, each science embodied for him different kinds of truths, not in opposition to or conflict with one another, but as complementary aspects of a grand project. Each field offered a different perspective, followed a slightly different method, and was seen to present a distinct face of reality, or at least reality in the only way that Man can perceive it. For Diderot, natural philosophy could not do without any of these facets; each was in some way crucial to the establishment of human knowledge. Thus we might encapsulate Diderot's conception of natural philosophy with the metaphor of a scientist peering at nature through the lens of a microscope that incorporates five filters. The microscope represents the intellectual apparatus of science, and although everything seen through the microscope is conditioned to some degree by the instrument itself, without it the reality being investigated would be totally inaccessible. Each of the filters could be seen as superimposed upon one another, a series of templates that conspire to produce a composite picture of the subject. Without each of these templates the vision offered by natural philosophy would be profoundly incomplete (perhaps even incomprehensible, since we could argue that they are all necessarily linked and in many ways interdependent). These filters are of course the five great scientific disciplines, which, in descending order, are mathematics, physics, chemistry, natural history and medicine. By descending order it is

4. Imre Lakatos, *The Methodology of scientific research programmes* (Cambridge, 1978).

simply meant here that mathematics is the most abstract discipline, the most removed from reality, while chemistry, natural history and medicine are all concerned with a much more concrete level of investigation. Of course, the farther we move down this ladder, the shakier the ground we walk on. Therefore, while mathematics offers the highest degree of certainty, it is also the most distant from its proper subject (i.e. the experience from which it has been induced), while the knowledge that pertains to the more experientially involved disciplines can only ever be categorised as probabilistic.

It might be further noted that there is something of a chronological relationship between the disciplines as well. Diderot's first love, scientifically speaking of course, was mathematics. This was followed in the 1740s by contemplation of questions pertaining to cosmology, the nature of life and the nature of the philosophical method itself. Although some of these questions were relevant to subjects such as medicine and natural history, Diderot articulated responses first in terms of general and experimental physics. Following hard on the heels of this phase of his thought came chemistry, which lent his experimental philosophy tremendous subtlety in terms of content. This, in turn, paved the way for a more integrated, more scientific, better substantiated approach to natural history and medicine, which in many ways served to crown Diderot's career. Not by chance the order of the disciplines as they are presented above are also the order in which they appear in Diderot's *Plan d'une université*. Students should start with mathematics, then move onto rational mechanics, experimental physics, natural history, and chemistry. Medicine, of course, was a separate faculty, but tellingly Diderot was most certainly of the opinion that no one could properly undertake the study of medicine without having become acquainted with at least a little bit of natural history and chemistry. Thus, this perception of the 'great disciplines' (as I have called them here) vis à vis Diderot serves a wide variety of purposes in terms of both ordering and understanding his natural philosophy; it is chronological, pedagogical, structural and philosophical.

Implicit in this theory of the various layers of Diderot's natural philosophical conception is a compelling philosophy of science. Much has already been implied here regarding this, but we might devote just a few thoughts to considering how it fits into the general scheme of the history of the philosophy of science. In modern parlance, Diderot would probably be called a 'normative naturalist', that is, one who views science and the philosophy of science as a seamless whole, and who is 'concerned to deny that the philosophy of science is a 'transcendent' discipline in which trans-historical, inviolable evaluative principles are superimposed upon the practice of science'.[5] The normative naturalist position posits

5. Losee, *A Historical introduction*, p.241.

that standards for the appraisal of scientific theories and explanations are, just like the very content of science itself, subject to perpetual correction and refinement. Thus, even for a realist such as Diderot, who accepted the existence of a world outside the minds of human beings, science, indeed all knowledge, is but part of an interminable process of trying to conceive of the world in an increasingly comprehensive, coherent and methodologically sound fashion. One of the most eloquent statements of such a position, and one that may well be applied directly to Diderot's world-view, is Otto Neurath's 'Boat' image of science. According to Neurath, the development of science was like the endless voyage of a ship on a boundless ocean, and 'we [human beings] are like sailors who have to rebuild their ship on the open sea, without ever being able to dismantle it in dry-dock and reconstruct it from the best components'.[6] Underlying this image are the assumptions that the ship can never attain the status of 'seaworthy for all time', that continual rebuilding is necessary, and that there is no objective standpoint from which to perform this rebuilding. As John Losee has claimed, Neurath's idea supposes that 'there is no inviolable axiological dimension within science [...] even the most general of evaluative principles are subject to modification in the course of the rebuilding process'.[7] Taking this brand of normative naturalism to be eminently applicable to Diderot's own ideas about natural philosophy, and bearing in mind all of the other methodological and epistemological lessons that he sought to teach both in *L'Intreprétation de la nature* and elsewhere, we may thus observe that Diderot falls within a line of philosophers that stretches from David Hume, to Ernst Mach, to Henri Poincaré and in some respects even to Einstein.

Having established then that Diderot was a serious, avid and comprehensive natural philosopher, as well as a philosopher of science, there are just two more things that need be said, both with reference to the implications of this study for future research on Diderot. First, more work needs to be done regarding the linkage between the nature of Diderot's physical sciences and his interest in the social sciences, or the sciences of man. A great deal of attention has been paid over the years to his politics, his aesthetics, his atheism and to a certain extent even his anthropology, but what is appropriate now would be a thorough examination of how these aspects of his thought are related – epistemologically, methodologically and philosophically – to the 'harder' sciences that have been under investigation in this book. There can be little doubt that it is well nigh impossible to extricate Diderot's interpretation of nature from his views regarding Man and society. Lines of connection have already been suggested here with regard to Diderot's

6. Otto Neurath, 'Protocol statements', in *Otto Neurath: philosophical papers*, ed. R. S. Cohen and M. Neurath (Dordrecht, 1983), p.92.

7. Losee, *A Historical introduction*, p.243.

burgeoning interest in probability theory, its applications to various realms of human affairs, and his enthusiasm for the potential of medicine to bring real improvements to the condition of his fellow men. Much more, however, remains to be explored. How did Diderot propose bringing about real change, and to what degree were his ideas about this conditioned by his vision of natural philosophy? How did his physiology really impact upon his view of morality, ethics, community and politics? To what degree can his views about abstraction, mathematics and the scientific genius be mingled with his aesthetic theory? To what degree is there a direct link between his scientific relativism and the cultural relativism manifested or hinted at in works such as *Supplément au Voyage de Bougainville*? And what of his views on women, children or 'natural man' versus 'civilised man'? All of these issues merit further research, and perhaps an investigation that parallels Keith Baker's effort on Condorcet, which traces that *philosophe*'s path from natural philosophy to social science.[8]

Finally, fresh scope has also been created here to examine more systematically the influence of Diderot's natural philosophical thought. Treatment of such matters is a notoriously difficult pursuit, but various scholars have been suggesting for many years that Diderot exerted some manner of influence on subsequent generations of thinkers.[9] Lamarck, Bichat, Bernard, Cabanis and the Ideologues, Goethe and even Lavoisier have all been mentioned, at one time or another, as potential benefactors of Diderot's philosophy. The obvious objection to following this up is that many of the *philosophe*'s most crucial works were not published until well after his death, and in some cases, well into the nineteenth century. Nevertheless, a number of pieces did circulate in manuscript form prior to official publication, and in this way may indeed have come into the relevant hands. Martin Staum has already pointed this out, and shown that Diderot's literary executor Naigeon passed a copy of *Le Rêve de D'Alembert* to close associates of Cabanis, creating the possibility that the ideologue did indeed read the work fairly early in his career.[10] Another possible medium for the transfer of ideas would, of course, have been the salons. For obvious reasons, it is difficult to appraise exactly what might have transpired in any given meeting, but a thorough examination of the relevant correspondence has the potential to shed some light on the issue. A final problem that besets any investigation regarding lines of intellectual influence is that one's subjects are not always candid about the nature or extent of their sources.

8. See Baker, *Condorcet*.

9. See, for instance, Mayer, *Diderot, homme de science*; Paitre, *Diderot biologiste*; P. H. Tribouillet, *Diderot et la médecine* (Lyon, 1921); Gerard Vasco, *Diderot and Goethe: a study in science and humanism* (Geneva, 1978); Staum, *Cabanis*; Mason, *The Irresistible Diderot*; Gillispie, 'The Encyclopédie and the Jacobin philosophy of science'.

10. Staum, *Cabanis*, p.91.

On account of all of these difficulties, any attempt to evaluate properly the impact of Diderot's natural philosophy on succeeding generations may ultimately be doomed to failure, or at the very least circumscribed within strict parameters. Nevertheless, there have been enough suggestions on the part of scholars to merit a thorough, systematic and scholarly examination of the issue. This kind of examination would undoubtedly rule out certain possibilities once and for all, but it might also open up interesting new doors, or lead to confirmation where once there was only speculation. Such an investigation has the potential to cement the re-evaluation of Diderot's thought that has been commenced here, and perhaps to raise or at least adjust his status both within the context of the Enlightenment and that of the history of science.

Appendix: chronological list of Diderot works referenced

Mémoires sur différents sujets de mathématiques, *OC*, vol.2 (1748)

Les Bijoux indiscrets, *OC*, vol.2 (1748)

Lettre d'un citoyen zélé, *OC*, vol.2 (1748/60)

Lettre sur les aveugles, *OC*, vol.2 (1749)

Prospectus de l'Encyclopédie, *OC*, vol.2 (1750)

Lettre sur les sourds et muets, *OC*, vol.2 (1751)

'Animal', in *Encyclopédie*, vol.1 (1751)

Suite de l'Apologie de l'abbé de Prades, *OC*, vol.2 (1752)

Pensées sur l'interprétation de la nature, *OC*, vol.2 (1753/54)

'Eclectisme', in *Encyclopédie*, vol.5 (1755)

'Encyclopédie', in *Encyclopédie*, vol.5 (1755)

Paris, B.N., n.a.fr. 4043-4044, *Cours de chimie, ou Leçons de Mr. Rouelle, recueilles pendant les années 1754, et 1755; Rédigées en 1756; Revues et corrigées en 1757 et 1758* (1757-1758)

Nouveaux mémoires sur différents sujets de mathématiques, *OC*, vol.5 (1761)

'Hypothèse', in *Encyclopédie*, vol.8 (1765)

'Naître', in *Encyclopédie*, vol.15 (1765)

Le Rêve de D'Alembert, *OC*, vol.8 (1769)

Principes philosophiques sur la matière et le mouvement, *OC*, vol.9 (1770)

Note historique sur Michel Vanloo, peintre, et le chimiste Rouelle, *OC*, vol.9 (1770)

Mathématiques, *OC*, vol.11 (1770s)

Expériences intéressantes, *OC*, vol.9 (1770s)

Eléments de physiologie, ed. Jean Mayer (Paris, 1964), (1770s)

Eléments de physiologie, ed. Paulo Quintili (Paris, 2002), (1770s)

Observations et questions sur la variole, *OC*, vol.9 (1774)

Réfutation d'Helvétius, *OC*, vol.11 (1774)

Refutation d'Hemsterhuis, *OC*, vol.11 (1774)

Voyage en Hollande et dans le Pays-Bas autrichien, *OC*, vol.11 (1774-1775)

Plan d'une université, *OC*, vol.11 (1775)
Le Neveu de Rameau, *OC*, vol.12 (1777-1780)
Jacques le fataliste, *OC*, vol.13 (1778-1780)
Essais sur les règnes de Claude et de Neron, *OC*, vol.13 (1782)

Bibliography

Alembert, Jean D', *Mélanges de littérature: d'histoire, et de philosophie* (Paris, Briasson, 1753).

–, *Œuvres de D'Alembert* (Paris, 1821).

Alexander, Peter, *Ideas, qualities and corpuscles: Locke and Boyle on the external world* (Cambridge, 1985).

Allan, D. G. C., and R. E. Schofield, *Stephen Hales, scientist and philanthropist* (London, 1980).

Anderson, Lorin, *Charles Bonnet and the order of the known* (Boston, MA, 1982).

Anderson, Wilda, *Diderot's dream* (Baltimore, MD, 1990).

Bacon, Francis, *Novum Organum*, ed. Lisa Jardine and Michael Silverthorne (Cambridge, 2000).

–, *Novum Organum*, in *Great books of the Western world*, ed. Robert Hutchins (Chicago, IL, 1952).

Baere, Benoît de, *Trois introductions à l'abbé Pluche: sa vie, son monde, ses livres* (Geneva, 2001).

Baker, Keith, *Condorcet: from natural philosophy to social mathematics* (Chicago, IL, 1975).

Barber, William, *Leibniz in France, from Arnauld to Voltaire: a study in French reactions to Leibnizianism, 1670-1760* (Oxford, 1955).

Barry Jr, James, *Measures of science* (Evanston, IL, 1996).

Beeson, David, *Maupertuis: an intellectual biography*, *SVEC* 299 (1992).

Belaval, Yvon, 'Le Matérialisme de Diderot', in *Europäische Aufklärung; Herbert Dieckmann zum 60 Geburtstag*, ed. H. Friedrich and F. Schaml (Munich, 1967).

Bensande-Vincent, Bernadette, and Isabelle Stengers, *A History of chemistry*, translated by Deborah van Dam (Cambridge, MA, 1996).

Black, J., and R. Porter (ed.), *A Dictionary of eighteenth-century history* (London, 2001).

Blum, Carol, *Diderot: the virtue of a philosopher* (New York, 1974).

Boas Hall, Marie, *Robert Boyle and seventeenth-century chemistry* (Cambridge, 1968).

Bordeu, Théophile de, *Théophile de Bordeu: Correspondance*, 3 vols, ed. Martha Fletcher *et al.* (Montpellier, 1977-1979).

–, 'Crise', in *Encyclopédie* 4 (Paris, Briasson, David, Le Breton et Durand, 1754).

Bouissounouse, Janine, *Julie de Lespinasse* (Paris, 1958).

Boullier, David-Renauld, *Essais philosophiques* (Amsterdam, 1728).

Boury, Dominique, 'Bordeu et Barthez: le malentendu vitaliste', in *La Lettre de la Maison Française d'Oxford*, no.14 (Oxford, 2001).

Bremner, Geoffrey, *Order and chance: the pattern of Diderot's thought* (Cambridge, 1983).

Brewer, Daniel, *The Discourse of Enlightenment in eighteenth-century France: Diderot and the art of philosophizing* (Cambridge, 1993).

Brockliss, Laurence, *French higher education in the seventeenth and eighteenth centuries: a cultural history* (Oxford, 1987).

–, *Calvet's web: Enlightenment and the Republic of Letters in eighteenth-century France* (Oxford, 2002).

–, and Colin Jones, *The Medical world of early modern France* (Oxford, 1997).

Brown, Gary, 'The Evolution of the term "mixed mathematics"', *Journal of the history of ideas*, 52: 1 (Jan.-Mar. 1991).

Brunet, Pierre, *L'Introduction des théories de Newton en France au XVIIIe siècle avant 1738* (Paris, 1931).

Bühler, W. K., *Gauss: a biographical study* (New York, 1981).

Burkhardt, Richard, 'The Inspiration of Lamarck's belief in evolution', in *Journal of the history of biology* 5 (1972).

–, *The spirit of system: Lamarck and evolutionary biology* (Cambridge, MA, 1977).

Cap, Paul-Antoine, *Guilliame-François Rouelle* (Paris, 1842).

Cassirer, Ernst, *The Philosophy of the Enlightenment*, translated by Fritz Koelln and James Pettegrove (Boston, 1955).

Chartier, Roger, *et al.*, *L'éducation en France du XVIe au XVIIIe siècle* (Paris, 1976).

Chevalier, Anne, *Jacques le fataliste* (Paris, 1993).

Chouillet, Jacques, *La formation des idées esthétiques de Diderot, 1745-1763* (Paris, 1973).

Cimino, Guido, and François Duchesneau, *Vitalisms: from Haller to the cell theory: proceedings of the Zaragoza Symposium: 19th international congress of history of science, 22-29 August 1993* (Florence, 1997).

Clarke, Desmond, *Descartes' philosophy of science* (Manchester, 1982).

Cohen, I. Bernard, *The Newtonian revolution* (Cambridge, 1980).

Cohen, R. S., and M. Neurath (ed.), *Otto Neurath: philosophical papers* (Dordrecht, 1983).

Conner, Clifford, *Jean Paul Marat: scientist and revolutionary* (Amherst, NY, 1999).

Constant, J. P., *L'Enseignement de la chimie au Jardin Royal des Plantes de Paris* (Cahors, 1952).

Coolidge, Julian Lowell, *The Mathematics of great amateurs* (Oxford, 1990).

Creech, James, *Diderot: thresholds of representation* (Columbus, OH, 1986).

Crocker, Lester, 'J. Toland et le materialisme de Diderot', in *Revue d'histoire littéraire de la France* 53 (1953) p.289-95.

–, *Diderot: the embattled philosopher* (New York, 1954).

–, 'Diderot and eighteenth-century French transformism', in *Forerunners of Darwin, 1745-1859*, ed. B. Glass, O. Temkin and W. Straus (Baltimore, MD, 1959).

–, 'The Idea of a neutral universe in the French Enlightenment', in *Diderot studies* 21 (1983).

Crosland, Maurice, 'Chemistry and the chemical revolution', in *The Ferment of knowledge*, ed. G. S. Rousseau and Roy Porter (London, 1980), p.389-416.

Culotta, Charles A., *Respiration and the Lavoisier tradition: theory and modification, 1777-1850* (Philadelphia, PA, 1972).

Cunningham, Andrew, and Roger French (ed.), *The Medical enlightenment of the eighteenth century* (Cambridge, 1990).

Daston, Lorraine, *Wonders and the order of nature, 1150-1750* (New York, 1998).

Dawson, Virginia, *Nature's enigma: the problem of the polyp in the letters of Bonnet, Trembley and Réaumur* (Philadelphia, PA, 1987).

Demoustier, A., and D. Julia (ed.), *The Jesuits, Ratio Studiorum: plan raisonné et institution des etudes dans la Compagnie de Jesus*, translated by L. Albrieux and D. Pralon-Julia (Paris, 1997).

Deparcieux, Antoine, *Essai sur les probabilités de le durée de la vie humaine* (Paris, Frères Guerin, 1746).

Descartes, René, *Treatise of Man*, translation and commentary by Thomas Steele Hall (Cambridge, MA, Harvard University Press, 1972).

Dhombres, Jean, 'Quelques rencontres de Diderot avec les mathématiques', in *Colloque internationale de Diderot*, ed. A.-M. Chouillet (1984).

Diderot, Denis, *Correspondance*, ed. Georges Roth (Paris, 1955-1970).

–, *Œuvres complètes (édition chronologique)*, 15 vols, ed. Roger Lewinter (Paris, 1963-1973).

–, *Eléments de physiologie*, ed. Jean Mayer (Paris, 1964).

–, *Eléments de physiologie*, ed. Paulo Quintili (Paris, 2002).

Dieckmann, Herbert, *Inventaire du Fonds Vandeul et Inédits de Diderot* (Geneva, 1951).

–, 'The First edition of Diderot's Pensées sur l'interprétation de la nature', in *Isis*, vol.46, part 3, no.145 (September 1955).

–, 'Observations sur les manuscrits de Diderot conservés en Russie', in *Diderot studies* 4 (1963).

Dilly, Antoine, *Traite de l'ame* (Paris, Gallet, 1691).

Dobbs, Betty Jo, *The Janus face of genius: the role of alchemy in Newton's thought* (Cambridge, 1991).

Doolittle, James, 'Robert James, Diderot, and the Encyclopédie', *Modern language notes* 71 (June 1956).

Dorveaux, P., 'Notes sur Mlle Biheron', in *La Médecine anecdotique, historique, littéraire* (Paris, Gallimard, 1900-1901).

Duchesneau, François, *La physiologie des lumières: empirisme, modèles et théories* (The Hague, 1982).

Duncan, Alistair, *Laws and order in eighteenth-century chemistry* (Oxford, 1996).

Ehrard, Jean, *L'Idée de nature en France dans la première moitié du XVIII*[e] *siècle*, vol.1 (Paris, 1963).

Ehrman, Esther, *Mme du Châtelet; scientist, philosopher and feminist of the Enlightenment* (Leamington Spa, 1986).

Encyclopédie, ou, Dictionnaire raisonné des sciences, des arts et des métiers (Paris, Briasson *et al.*, 1751-1765).

Fara, Patricia, *An Entertainment for angels: electricity in the Enlightenment* (Cambridge, 2002).

Fischer, J.-L., 'L'Encyclopédie présente-t-elle une pré-science des monstres?', in *Recherche sur Diderot et l'Encyclopédie* 16 (1994), p.133-52.

Forno, Laurence, *The cosmic mysticism of Diderot*, *SVEC* 143 (1975), p.113-40.

Foucault, Michel, *Archaeology of knowledge* (London, 1972).

France, Peter, *Diderot* (Oxford, 1983).

Frängsmyr, Tore, *Linnaeus, the man and his work* (Canton, MA, 1994).

–, *et al.*, *The quantifying spirit in the 18th century* (Berkeley, CA, 1990).

Furbank, Philip, *Diderot* (London, 1993).

Gerbi, Antonello, *The Dispute of the new world: the history of a polemic, 1750-1900*, translated by Jeremy Moyle (Pittsburgh, PA, 1973).

Gerhardt, C. J. (ed.), *Die philosophischen Schriften von Gottfried Wilhelm Leibniz*, 7 vols (Berlin, 1875-1890).

Gillispie, Charles Coulston, 'The Encyclopédie and the Jacobin philosophy of science: a study in ideas and consequences', in *Critical problems in the history of science*, ed. Marshall Clagett (Madison, WI, 1962), p.255-89.

–, *Genesis and geology: a study in the relations of scientific thought, natural theology, and social opinion in Great Britain, 1790-1850* (New York, 1959).

Guerlac, Henry, *Lavoisier – the crucial year: the background and origin of his first experiments on combustion in 1772* (New York, 1961).

–, *Newton on the Continent* (Ithaca, NY, 1981).

Guiragossian Carr, Diana, 'The Image of Diderot in the British periodical press from 1750-1800', in *Diderot: digression and dispersion*, ed. Jack Undank and Joseph Herbert (Lexington, KY, 1984).

Haigh, Elizabeth, *Xavier Bichat and the medical theory of the eighteenth century* (London, 1984).

Hales, Stephen, *Vegetable staticks*, foreword by M. A. Hoskin (London, 1961, first published 1727).

Hall, A. Rupert, *All was light: an introduction to Newton's Opticks* (Oxford, 1993).

Hall, Thomas S., *Ideas of life and matter*, 2 vols (Chicago, IL, University of Chicago Press, 1969).

Hankins, Thomas, *Jean D'Alembert; science and the Enlightenment* (Oxford, 1970).

Hanna, Blake, 'Diderot théologien', in *Revue d'histoire littéraire de France*, vol.78 (1978).

Hemsterhuis, François, *Lettre sur l'homme et ses rapports, avec le commentaire inédit de Diderot*, ed. Georges May (New Haven, CT, 1964).

Henry, Charles, 'Introduction a la chymie; manuscrit inédit de Diderot', in *Revue scientifique*, 2e semestre (3e serie), no.4, 26 July 1884.

Henry, John, *Knowledge is power; how magic, the government and an apocalyptic vision inspired Francis Bacon to create modern science* (Cambridge, 2002).

Hill, Emita, 'Materialism and monsters in *Le Rêve de D'Alembert*', in *Diderot studies* 10 (1968).

Hobson, Marian, '"Nexus effectivus" and "nexus finalis"', in *Rousseau and the eighteenth century*, ed. M. Hobson, J. Leigh and R. Wokler (Oxford, 1992).

–, and Simon Harvey (ed.), *Lettre sur les aveugles: à l'usage de ceux qui voient ; Lettre sur les sourds et muets, à l'usage de ceux qui entendent et qui parlent* (Paris, 2000).

Holmes, Frederic Lawrence, *Eighteenth-century chemistry as an investigative enterprise* (Berkeley, CA, 1989).

–, *Antoine Lavoisier – the next crucial year* (Princeton, NJ, 1997)

Israel, Jonathan, *Radical Enlightenment: philosophy and the making of modernity 1650-1750* (Oxford, 2002).

Jacques, Jean, 'Le Cours de chimie de G.-F. Rouelle recueilli par Diderot', in *Revue d'histoire des sciences* 38:1 (Jan.-Mar. 1985) p.43-54.

Jaffe, Bernard, *Crucibles: the story of chemistry from ancient alchemy to nuclear fission* (New York, 1998).

Jullian, Vincent, *Descartes, La Géométrie de 1637* (Paris, 1996).

Kafker, Frank A., *The Encyclopedists as a group: a collective biography of the authors of the 'Encyclopédie'*, *SVEC* 345 (1996).

Kaitaro, Timo, *Diderot's holism: philosophical anti-reductionism and its medical background* (Frankfurt, 1997).

Kell, Patricia, *British collecting 1656-1800: scientific enquiry and social practice* (D.Phil. thesis, Oxford, 1996).

Kemp, Jonathan, *Diderot: interpreter of nature* (New York, 1943).

Kiernan, Colm, 'Additional reflections on Diderot and science', in *Diderot studies* 14 (1971).

Knight, Isabel, *The Geometric spirit* (New Haven, CT, 1968).

Kors, Alan, *D'Holbach's coterie: an enlightenment in Paris* (Princeton, NJ, 1976).

Krakeur, Lester, and Raymond Kruger, 'The Mathematical writings of Diderot', in *Isis*, vol.33, part 2, no.88 (June 1941).

Kuhn, Thomas, *The Essential tension; selected studies in scientific tradition and change* (Chicago, IL, 1977).

Lakatos, Imre, *The Methodology of scientific research programmes* (Cambridge, 1978).

Lakoff, George, and Rafael E. Nez, *Where mathematics comes from: how the embodied mind brings mathematics into being* (New York, 2000).

Licoppe, Christian, *La formation de la practique scientific: le discours de l'expérience en France et en Angleterre (1630-1820)* (Paris, 1996).

Lindeboom, Gerrit Arie, *Hermann Boerhaave: the man and his work* (London, 1968).

Liublinskii, Vladimir Sergeevich, 'Sur

les traces des livres lus par Diderot', *Europe*, 405-406 (Jan.-Feb. 1963).

Losee, John, *A Historical introduction to the philosophy of science*, 4th edn (Oxford, 2001).

Lough, John, *The Encyclopédie* (London, 1971).

Louis, Antoine, 'Chirurgie', in *Encyclopédie* 3 (Paris, Briasson, David, Le Breton et Durand, 1753).

Loy, John, *Diderot's determined fatalist: a critical appreciation of Jacques le fataliste* (New York, 1950).

Luxembourg, Lilo K., *Francis Bacon and Denis Diderot: philosophers of science* (Copenhagen, 1967).

Marchal, Roger, *Fontenelle à l'aube des Lumières* (Paris, 1997).

Martin, Julian, 'Sauvages' nosology: medical enlightenment in Montpellier', in *The Medical enlightenment of the eighteenth century*, ed. Andrew Cunningham and Roger French (Cambridge, 1990), p.111-37.

Mason, John Hope, *The Irresistible Diderot* (London, 1982).

Maupertuis, Pierre-Louis Moreau de, *Œuvres de Maupertuis*, ed. Jean-Marie Bruyset (Lyon, Jean-Marie Bruyset Imprimeur-Libraire, 1758).

Mayer, Jean, *Diderot, homme de science* (Rennes, 1959).

–, 'Cours de chimie de Rouelle', in *Diderot: œuvres complètes*, vol.9 (Paris, 1981).

McCann, H. Gilman, *Chemistry transformed: the paradigmatic shift from phlogiston to oxygen* (Norwood, NJ, 1978).

McNeill, Maureen, *Under the banner of science: Erasmus Darwin and his age* (Manchester, 1987).

Melhado, Evan, 'Chemistry, physics, and the chemical revolution', in *Isis* vol.76 (1985), p.195-211.

Mémoires de l'Académie des sciences (Paris, 1744, 1745 and 1754).

Mercier, Louis-Sebastien, *Tableau de Paris*, 12 vols (Amsterdam, [s.n.], 1782-1788).

Metzger, Hélène, *Newton, Stahl, Boerhaave et la doctrine chimique* (Paris, 1974).

Miller, G., *The Adoption of inoculation for smallpox in England and France* (Philadelphia, PA, 1957).

Morgan, Augustus de, *A Budget of paradoxes* (London, 1872).

Morilhat, Claude, *La Mettrie: un matérialisme radical* (Paris, 1997).

Morin, Robert, 'Diderot et l'imagination', in *Annales littéraires de l'Université de Besançon* (Paris, 1987).

Naville, Pierre, *D'Holbach et la philosophie scientifique au XVIIIe siècle* (Paris, 1967).

Newton, Isaac, *Opticks: or treatise of the reflections, refractions, inflections and colours of light* (New York, 1952).

–, *Principia mathematica*, translated by Andrew Motte, edited and presented by Florian Cajori (Berkeley, CA, 1960).

Paitre, Fernand, *Diderot biologiste* (Lyon, 1904).

Pappas, John, *L'Esprit de finesse contre l'esprit de géométrie: un debate entre Diderot et d'Alembert*, *SVEC* 89, 1972.

Partington, J. R., *A History of chemistry*, vol.3 (London, 1962).

Pearson, Roger, *Voltaire almighty: a life in pursuit of freedom* (London, 2005).

Peltonen, Markku (ed.), *The Cambridge companion to Bacon* (Cambridge, 1996).

Pérez-Ramos, Antonio, *Francis Bacon's idea of science and the maker's knowledge tradition* (Oxford, 1988).

Perkins, J. A., *Diderot et La Mettrie*, *SVEC* 10 (1959).

Poincaré, Henri, *Science and hypothesis*, translated by G. B. Halsted (New York, 1905).

Polinière, Pierre, *Expériences de physique* (Paris, Jean de Laulne *et al*, 1741).

Principe, Lawrence, *The Aspiring adept: Robert Boyle and his alchemical quest* (Princeton, NJ, 1998).

Proust, Jacques, 'La Bibliothèque de

Diderot', in *Revue des sciences humaines* 33 (1958).
–, *Diderot et l'Encyclopédie* (Paris, 1962).

Rappaport, Rhoda, *G.-F. Rouelle: his cours de chimie and their significance for eighteenth century chemistry* (Doctoral thesis, Cornell University, NY, 1958).
–, 'G.-F. Rouelle: an eighteenth-century chemist and teacher', in *Chymia*, vol.6 (1960), p.68-101.
–, 'Rouelle and Stahl – the phlogistic revolution in France', in *Chymia*, vol.7 (1961), p.73-102.
–, 'Geology and orthodoxy: the case of Noah's flood in eighteenth-century thought', in *The British journal for the history of science* (March 1978).
Raymond, Catherine, 'Diderot et la médecine', in *Litterature et médecine II*, ed. Jean-Louis Cabanes, *Cahiers du laboratoire pluridisciplinaire de Recherches sur l'imaginaire appliquées à la litterature*, no.55 (July 2000).
Resnik, Michael, *Frege and the philosophy of mathematics* (Ithaca, NY, 1980).
Rey, Roselyne, 'Dynamique des formes et interprétation de la nature', in *Recherches sur Diderot et l'Encyclopédie*, vol.11 (1991).
–, *Naissance et développement du vitalisme en France de la deuxième moitié du 18e siècle à la fin du Premier Empire*, *SVEC* 381 (2000).
Riley, James C., *The Eighteenth-century campaign to avoid disease* (London, 1987).
Roe, Shirley, *Matter, life, and generation: eighteenth-century embryology and the Haller–Wolff debate* (Cambridge, 1981).
Roger, Jacques, 'Diderot et Buffon en 1749', in *Diderot studies* 4 (1963).
–, 'The living world', in *The Ferment of knowledge*, ed. G. S. Rousseau and Roy Porter (Cambridge, 1980).
–, *Buffon: a life in natural history*, translated by Sarah Lucille Bonnefoi (Ithaca, NY, 1997).
–, *The Life sciences in eighteenth-century French thought*, ed. Keith R. Benson, translated by Robert Ellrich (Stanford, CA, Stanford University Press, 1997).
Rousseau, G. S., and Roy Porter (ed.), *The Ferment of knowledge* (Cambridge, 1980).
Rudwick, Martin, *The Meaning of fossils: episodes in the history of palaeontology* (New York, 1976).

Schlegel, Dorothy, *Shaftesbury and the French deists* (Chapel Hill, NC, 1956).
Schubert-Soldern, Rainer, *Mechanism and vitalism: philosophical aspects of biology* (Notre Dame, IN, 1962).
Shapin, Steven, *Leviathan and the air-pump: Hobbes, Boyle, and the experimental life* (Princeton, NJ, 1985).
Sheynin, O. B., 'Daniel Bernoulli's work on probability', in *Studies in the history of statistics and probability*, ed. M. G. Kendall and R. L. Plackett (London, 1977), p.105-33.
Sloan, Phillip, 'Natural history, 1670-1802', in *Companion to the history of modern science*, ed. R. C. Olby *et al* (London, 1990).
Spallanzani, Lazzaro, *Nouvelles recherches sur les découvertes microscopiques, et la génération des corps organisés. Ouvrage tr. par l'abbé Regley. Avec des notes, des recherches physiques & métaphysiques sur la nature & la religion [&c.] par m. de Needham* (London and Paris, Lacombe, 1769).
Spary, Emma, *Utopia's garden: French natural history from old regime to revolution* (Chicago, IL, 2000).
Stafleu, A., *Linnaeus and the Linnaeans: the spreading of their ideas in systematic botany, 1735-1789* (Utrecht, 1971).
Staum, Martin, *Cabanis: Enlightenment and medical philosophy in the French Revolution* (Princeton, NJ, 1980).
Steinke, Hubert, 'Haller's concept of irritability and sensibility and its reception in france', in *La Lettre de la Maison Française d'Oxford*, no.14 (Oxford, 2001), p.37-69.

Stevens, Peter, *The development of biological systematics: Antoine-Laurent de Jussieu, nature, and the natural system* (New York, 1994).

Strathern, Paul, *Mendeleyev's dream: the quest for the elements* (London, 2001).

Terrall, Mary, *The Man who flattened the earth: Maupertuis and the sciences in the Enlightenment* (Chicago, IL, 2002).

Thomson, Ann, 'L'Unité matérielle de l'homme chez La Mettrie et Diderot', in *Colloque international Diderot* (Paris, 1985), p.61-68.

Toland, John, *Letters to Serena* (London, printed for Bernard Lintot, 1704).

Torlais, Jean, *L'abbé Nollet: un physicien au siècle des Lumières* (Argueil, 1987).

Tribouillet, P. H., *Diderot et la médecine* (Lyon, 1921).

Undank, Jack, and Herbert Josephs (ed.) *Diderot: digression and dispersion: a bicentennial tribute* (Lexington, KY, 1984).

Vartanian, Aram, 'From deist to atheist: Diderot's philosophical orientation 1746-1749', in *Diderot studies* 1 (1949).

–, *Diderot and Descartes: a study of scientific naturalism in the Enlightenment* (Princeton, NJ, 1953).

–, *L'Homme machine: a study in the origins of an idea* (Princeton, NJ, 1960).

–, 'The Problem of generation and the French Enlightenment', in *Diderot studies* 6 (1964).

–, *Science and humanism in the French Enlightenment* (Charlottesville, VA, 1999).

Vasco, Gerard, *Diderot and Goethe: a study in science and humanism* (Geneva, 1978).

Venturi, Franco, *La jeunesse de Diderot, 1713-1753*, translated by Juliette Bertrand (Paris, 1939).

Wartofsky, Max, 'Diderot and the development of materialist monism', in *Diderot studies* 2 (1952).

Wellmann, Kathleen Anne, *La Mettrie: medicine, philosophy, and Enlightenment* (Durham, NC, 1992).

White, John Henry, *The History of the phlogiston theory* (London, 1932).

Williams, Elizabeth, *The Physical and the moral: anthropology, physiology, and philosophical medicine in France, 1750-1850* (Cambridge, 1994).

Wilson, Arthur, 'Leningrad, 1957: Diderot and Voltaire gleanings', *French review* 31 (1958).

–, *Diderot* (New York, 1972).

Winter, Ursula, *Der Materialismus bei Diderot* (Geneva, 1972).

Yoder, Joella G., *Unrolling time: Christiaan Huygens and the mathematization of nature* (Cambridge, 1988).

Yolton, John W., *Thinking matter: materialism in eighteenth-century Britain* (Oxford, 1984)

–, *Locke and French materialism* (Oxford, 1991).

– (ed.), *Philosophy, religion and science in the seventeenth and eighteenth centuries* (Rochester, NY, 1990).

Index